AGRICULTURAL RESEARCH UPDATES

RAW MILK

PRODUCTION, CONSUMPTION AND HEALTH EFFECTS

AGRICULTURAL RESEARCH UPDATES

Additional books in this series can be found on Nova's website under the Series tab.

Additional E-books in this series can be found on Nova's website under the E-book tab.

FOOD AND BEVERAGE CONSUMPTION AND HEALTH

Additional books in this series can be found on Nova's website under the Series tab.

Additional E-books in this series can be found on Nova's website under the E-book tab.

AGRICULTURAL RESEARCH UPDATES

RAW MILK

PRODUCTION, CONSUMPTION AND HEALTH EFFECTS

JANA MOMANI
AND
AHMAD NATSHEH
EDITORS

NOVA

Nova Science Publishers, Inc.
New York

For permission to use material from this book please contact us:
Telephone 631-231-7269; Fax 631-231-8175
Web Site: http://www.novapublishers.com

NOTICE TO THE READER

Additional color graphics may be available in the e-book version of this book.

LIBRARY OF CONGRESS CATALOGING-IN-PUBLICATION DATA

Raw milk : production, consumption and health effects / editors: Jana Momani and Ahmad Natsheh.
p. cm.
Includes index.
ISBN 978-1-61470-641-0 (hardcover)
1. Raw milk. 2. Milk yield. 3. Milk consumption. 4. Milk--Health aspects. I. Momani, Jana. II. Natsheh, Ahmad.
SF251.R39 2011
637'.1--dc23

2011025520

Published by Nova Science Publishers, Inc. † New York

CONTENTS

PREFACE

In this book, the authors gather topical research in the study of the production, consumption and health effects of raw milk. Topics discussed in this compilation include the recent facts on spoilage organisms and enzymes of microbial origin and their importance through the dairy chain; the identification of lipolytic and/or proteolytic psychotrophic Pseudomonas species in raw milk; raw sheep milk consumption and health effects in the province of Karak, Jordan and camel milk as a therapeutic alternative to treat diabetes.

Chapter 1 - Bacterial spoilage of milk and dairy products causes great economical losses for the dairy industry. This chapter reviews current knowledge on the most important spoilage organisms and enzymes of microbial origin and their importance throughout the dairy chain in light of commercially applied processing conditions. The organoleptic and texture effects of spoilage enzymes on milk and dairy products are also discussed. Aerobe spore-formers belonging to the genus *Bacillus sensu lato* and psychrotolerant Gram-negative rods belonging to the genus *Pseudomonas* are considered the most important spoilage micro-organisms in dairy products. Furthermore, the former do not only affect the quality of dairy products but are also occasionally implicated in food intoxications. Operational management throughout the dairy chain can influence species composition and bacterial load of raw milk prior to processing. At the farm, variable feeding and housing strategies of cows, as well as seasonal differences, can influence the microbial quality of milk. Furthermore, psychrotolerant bacteria, such as the pseudomonads, will benefit from prolonged cold storage throughout the dairy chain. Though these spoilage organisms have been subject of many studies and are thus historically well-known, recent large-scale raw milk isolation campaigns with identification based on current taxonomic insights and coupled to an extensive screening for enzymatic properties, support the need for re-evaluating the dominant species concerning dairy spoilage within these two groups of organisms (*Bacillus s.l.* and the genus *Pseudomonas*).

Chapter 2 - Many types of microorganisms are present in the milk collection environment and diversity in the raw milk microflora is typical, without dominance of a single species. The proportion of psychrotrophic bacteria in raw milk can vary widely and is associated with the level of farm hygiene. Studies in Europe have shown that typically, no more than 10% of the flora of good quality milk will be psychrotrophic with *Pseudomonas* species comprising a substantial proportion of these. *Pseudomonas fluorescens*, the most common species of the genus present in raw milk, has been involved in bacterial spikes (sudden elevations in total bacterial count) in farm bulk tank milk. Psychrotrophic *Pseudomonas* species play an important role in spoilage of UHT milk through the production of heat-stable lipases and proteases in raw milk that retain activity following UHT processing. Lipase and protease, produced by psychrotrophic *Pseudomonas* species are detected when the cell count exceeds $\sim 10^6$ cfu/mL. Prolonged refrigerated (4 °C) storage of raw milk increases the proportion of *Pseudomonas* species as do slightly higher temperatures (for example 6 °C) over a shorter period of time. This in turn increases the likelihood that they will produce heat-stable lipases and proteases. Furthermore, temperature fluctuations have been shown historically to occur in farm bulk milk, and the temperature of raw milk at the time of collection can vary widely. While less likely to occur today, both these scenarios could further compound the problem of *Pseudomonas* species proliferation in raw milk. The aim of the present study was to investigate the use of pulsed field gel electrophoresis (PFGE) for identifying sources of lipase and/or protease producing psychrotrophic *Pseudomonas* species at various pre-processing locations, and to track the types identified through the pre-processing environment. Incubation of raw milk was also carried out to simulate possible scenarios where the raw milk may be stored on the farm and in the silo prior to UHT processing. This enabled enrichment for spoilage bacteria and studies to identify sources of microorganisms that may contribute to lipolysis and proteolysis in raw and, subsequently, UHT milk or other long life dairy products. The impact of various storage conditions on the different Pulsed Field (PF) types of importance with regard to lipase and protease production was also assessed.

Chapter 3 - Sheep milk characterized by its high percentage of fat (6-8%) and high protein percentage (4.2-4.8), besides it has a very pronounce organoleptic characteristics which make it ideal to produce dairy products with a very special taste and with long shelf-life (ghee, Jameed and Baladi cheese). This article showed that a deficient milk refrigeration system in the small farm, beside the lack of sanitation during milking and handling constitute major factors in milk deterioration. Pasteurization of Baladi cheese milk and the boiling process of Baladi cheese have a great effort on improving the microbiological quality and the sensory evaluation of the final product.

Chapter 4 - The milk production has been growing around the world, but the biggest growth is in South and North America (Brazil and USA) and Asia (India and China). World cow's milk production in 2008 stood at over 578 million tones, with the top ten producing countries representing about 55.4% of production. Countries with advantage on land and animal feed will be a differential of productivity, such as India, China and Brazil. The consumption has grown following the increase in population and income. The countries from North America and Oceania are the biggest consumer, but don't consume the needs, which is about 200 liters per capita per year (WHO). The lowest consume is observed in countries from Asia and Africa, but just in this countries are observed the biggest growth in income. The quantity of milk's ingestion must be considered, since the vitamins and supplements are necessary to bones, muscles and immune system. Health benefits of milk included good bone health, robust skin, good immune system, prevention of illnesses such as hypertension, dental decay, dehydration, respiratory problems, obesity, osteoporosis and even some forms of cancer. The beneficial health nutrients obtained from milk are mandatory for human body and help in prevention of chronic ailments. Keeping away severe illnesses and harmful factors can be done through increasing milk consumption.

Chapter 5 - This study was performed to evaluate the efficacy of camel milk on alloxan-induced diabetic dogs and to follow this effect in addition to Can-insulin®. Four groups, composed of 4 diabetic dogs each, were used as follow: group 1 was getting camel milk, and group 2 treated simultaneous with camel milk and Can-insulin®, and group 3 received cow milk simultaneous with Can-insulin®. Group 4 contained clinically healthy animals and was used as control. Each dog received 500 ml of milk/day during five weeks. After three weeks, group 1 showed a significant decline on blood glucose levels from 10.33 ± 0.55 to 6.22 ± 0.5 mmol/L, this improvement on glycemic control was accompanied to a significant decrease on total proteins concentrations (from 79.66 ± 2.11 to 63.63 ± 4.43 g/L). A significant decline of cholesterol levels (from 6.84 ±1.2 to 4.9 ± 0.5 mmol/L) was shown after only two weeks of treatment. The same result was illustrated on group 2 treated simultaneous with camel milk and Can-Insulin. In group 3 the effect of Can-insulin was well shown only on blood glucose levels during the treatment. The investigation in this research was the beneficial effect of camel milk on diabetic dogs and its independence to the treatment with Can-insulin®.

Chapter 6 - Milk is a type of nutritionally complete food which contains protein, fat, lactose, vitamins, and minerals. The high nutritional content value of milk has become an excellent broth for a variety of microorganisms, which include many sorts of pathogens, such as (*Escherichia. coli, Listeria), (monocytogenes* and *Bacillus cereus*); (Fox and Cameron, 1982). The main purpose of pasteurization is to exterminate such pathogens in order to ensure the safety of milk and extend its shelf life. However, the pasteurization could also influence the physicochemical properties of milk, such as the changes of nutrient component which may reduce the digestibility and nutritional value of milk. Meantime, the sensory quality of milk also decreased slightly due to the heat treatment.

Chapter 7 - On one hand, according to FAO about 80% of the milk consumed worldwide is mostly obtained out of standards; in developed countries on the other hand an effective cold chain selects for spoiling bacteria that inflict significant losses to the dairy industry. Most studies, that concern modified or controlled atmospheres applied to bovine raw milk, were mostly based on CO_2 treatments, or for a few on mixtures of CO_2 and N_2 gases; a commonly accepted thought is that antimicrobial effects are associated with the application of CO_2, whereas N_2 has been employed as an inert gas component. Some recent studies, performed with an open system, based on a constant flushing of N_2 gas through the headspace of a vessel, at laboratory or at pilot scale suggest that bacterial growth could be substantially reduced by flushing pure N_2 gas alone into raw milk, with significant effects on mesophilic and psychrotrophic aerobes, but also on some other bacterial groups, without favouring the growth of anaerobes. One major observation was that phospholipases producers among them *Bacillus cereus* could be excluded at laboratory scale by the N_2 gas-based flushing; the inhibitory effect was also noticeable to some extend at pilot scale. Possible antimicrobial mechanisms underlying the use of N_2 gas, as well as the potential of controlled atmospheres-based treatments of raw milk will be discussed.

In: Raw Milk
Editors: J. Momani, et al
ISBN: 978-1-61470-641-0

Chapter 1

MICROBIAL CONTAMINATION AND SPOILAGE OF CONSUMER MILK – FACTS AND FICTION

***Valerie De Jonghe*[1], *An Coorevits*[2,3], *Sophie Marchand*[1], *Anita Van Landschoot*[2], *Jan De Block*[1], *Els Van Coillie*[1], *Paul De Vos*[3] *and Marc Heyndrickx*[1,4]**

[1]Institute for Agricultural and Fisheries Research (ILVO), Technology and Food Science Unit, Brusselsesteenweg 370, 9090 Melle, Belgium.
[2]Laboratory of Biochemistry and Brewing, Faculty of Applied Engineering Sciences, University College Ghent, Campus Schoonmeersen, Schoonmeersstraat 52, 9000 Ghent, Belgium.
[3]Laboratory of Microbiology (LM-UGent), Department of Biochemistry and Microbiology, Faculty of Sciences, Ghent University, K.L. Ledeganckstraat 35, 9000 Ghent, Belgium
[4]Department of Pathology, Bacteriology and Poultry Diseases, Faculty of Veterinary Sciences, Ghent University, Salisburylaan, Merelbeke

ABSTRACT

Bacterial spoilage of milk and dairy products causes great economical losses for the dairy industry. This chapter reviews current knowledge on the most important spoilage organisms and enzymes of microbial origin and their importance throughout the dairy chain in light of commercially applied processing conditions. The organoleptic and texture effects of spoilage enzymes on milk and dairy products are also discussed.

Aerobe spore-formers belonging to the genus *Bacillus sensu lato* and psychrotolerant Gram-negative rods belonging to the genus *Pseudomonas* are considered the most important spoilage micro-organisms in dairy products. Furthermore, the former do not only affect the quality of dairy products but are also occasionally implicated in food intoxications.

Operational management throughout the dairy chain can influence species composition and bacterial load of raw milk prior to processing. At the farm, variable feeding and housing strategies of cows, as well as seasonal differences, can influence the microbial quality of milk. Furthermore, psychrotolerant bacteria, such as the pseudomonads, will benefit from prolonged cold storage throughout the dairy chain.

Though these spoilage organisms have been subject of many studies and are thus historically well-known, recent large-scale raw milk isolation campaigns with identification based on current taxonomic insights and coupled to an extensive screening for enzymatic properties, support the need for re-evaluating the dominant species concerning dairy spoilage within these two groups of organisms (*Bacillus s.l.* and the genus *Pseudomonas*).

MILK: BORN TO BE SPOILED

The different constituents of milk make it a desired target for spoilage. This spoilage can either have an indigenous nature, or it can be attributed to microbial contamination.

Raw milk mainly consists of water (87%), carbohydrates (mainly lactose) (4.9%), lipids (3.7%), proteins (3.5%; mainly caseins and whey proteins), and minerals (0.7%) (Mabbit 1981). These percentages represent average values since the biochemical composition of milk varies according to different parameters: breed, age and feed of the cow and the stage of lactation (Verdier-Metz et al. 2009). Especially the fat content is highly susceptible to variations, whereas the amount of lactose remains more or less the same during the day and the different stages of lactation.

Though its structure appears to be homogenous, milk is composed of five physical phases: (i) casein micelles, (ii) fat globules, (iii) milk cells (commonly referred to as somatic cells, consisting predominantly of excreted epithelial cells and leukocytes which serve as a defense against pathogens), (iv) milk serum lipoprotein membrane (MSLM) vesicles (comprising 40-60% of the membranous phospholipids, the remainder being associated with the milkfat globule membrane (MFGM)) and (v) whey (milk serum) in which all other phases are homogenously dispersed (Silanikove et al. 2006;2008).

Protein Content

Caseins are the most important milk proteins, representing 76-86% of the total amount of proteins in cow's milk (Mabbit 1981). Eighty to ninety-five percent of all casein in normal milk is organized into casein micelles, spherical structures with a diameter ranging in size from 50-500 nm. Various models are proposed that describe the casein micelle structure (Phadungath 2005): the most widely accepted subunit model from Walstra (1999) states that casein micelles consist of a complex of sub-micelles, that are themselves built up of a hydrophobic core consisting of α- and β-caseins and a hydrophilic coat of κ-caseins (Figure 1). The hydrophilic parts of κ-casein contain carbohydrate groups, which project from the outsides of the complex micelles thus stabilizing the micelles against aggregation.

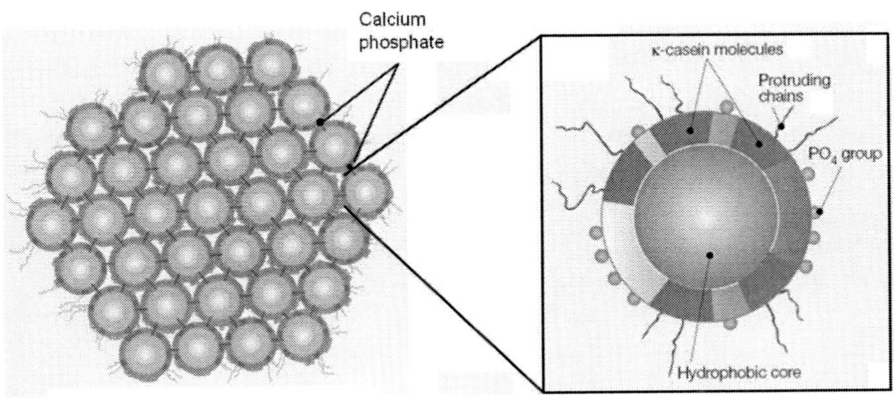

Figure 1. Sub-micelle model of the casein micelle as proposed by Walstra (1999). Adapted from Dairy Processing Handbook (adapted from Bylund, 1995).

The degradation of milk proteins, mainly caseins, through proteolysis may have beneficial effects and even be essential to obtain desirable qualities in food products, as is the case for flavour development and texture changes during cheese ripening. However, uncontrolled or unwanted proteolysis can adversely affect food quality: proteases are known to cause off-flavours because of the formation of 'bitter peptides'. These are small peptides that often contain high proportions of leucine, valine and aromatic amino acid residues, although bitterness is shown to be related to the hydrophobicity of casein-derived peptides rather than to specific amino acid residues or chain length (Ney 1979). Even though bacterial proteases can have substantial activity at low temperatures and at the pH of milk (pH 6.7), they do not often cause noticeable off-flavours in pasteurized milk. This may be explained by the short storage period that does not allow more advanced proteolysis which is required to obtain these small peptides (Mottar 1989). The shelf life of UHT-treated milk on the other hand, seems to be mainly limited by the action of heat resistant proteases during storage (Mottar et al. 1979): at first, a bitter flavour may occur (McKellar et al. 1984), and finally the deterioration can lead to gelation (Law et al. 1977) caused by formation of a three-dimensional matrix of aggregated β-lactoglobulin-κ-casein-complexes (Datta and Deeth 2001).

Proteases from bacterial origin may have multiple effects on cheese production. Loss of cheese yield by breakdown of casein is usually associated with increased storage time of the milk and a high psychrotolerant count (Mottar 1989;Yan et al. 1983). Cheese quality can be affected during storage by the action of bacterial proteases, resulting in an increased growth of starter cultures due to greater accessibility of nitrogen sources; however, this effect is rather minor since these enzymes are usually removed in the whey during cheese production - unlike bacterial lipases that are concentrated along with the fat in the curd (Fox 1981). Texture problems have also been associated with proteolysis, but only with milk with a high bacterial count before pasteurization (Law 1979). Problems with the quality of fermented milk products due to proteolytic activity have rarely been reported, probably due to their high acidity and low storage temperature below 10°C (Law 1979).

Different enzymes can be responsible for proteolytic decay: indigenous proteolytic enzymes and proteases from microbial origin. The effect of the two protease types in UHT milk is quite distinct: bacterial proteases lead to the formation of a curd or a gel with custard-like consistency throughout the whole milk sample (Hardham 1998), while the native plasmin causes a creamy layer on the surface of the milk which eventually thickens to form a curd-like layer (Harwalker 1982). Gels caused by bacterial proteases have a tighter protein network with thicker strands and contain more intact casein micelles and micelle aggregates than plasmin-initiated gels (Fox 1981;Harwalker 1982). Plasmin and bacterial proteases also show different affinities for the individual caseins: in contrast to plasmin, bacterial proteases have a preference for the hydrophilic κ-casein fraction that is readily available at the surface of the casein micelle followed by extensive non-specific hydrolysis (Cousin

1989;Guinot-Thomas et al. 1995). From the published data, it can be concluded that the order of susceptibility of the caseins to hydrolysis by bacterial proteases and plasmin are $\kappa>\beta>\alpha_{s1}$ and $\beta=\alpha_{s2}>\alpha_{s1}>\kappa$, respectively (Datta and Deeth 2001;Law 1979). However, when milk is cooled to 4°C, the casein micelle dissociates, increasing the amount of soluble casein from 15 to 30% (McMahon and Brown 1984), making milk altogether more susceptible to proteolysis. Furthermore, as bacterial proteases may also act as plasminogen activators (Figure 2) (Kohlmann et al. 1991) and/or disrupt the casein micelle causing release of plasmin into the milk serum (Fajardo-Lira et al. 2000), the relative significance of plasmin and bacterial proteases in age gelation of UHT stored milk is somewhat blurred.

Plasmin, the major indigenous protease in milk, is part of a complex known as the plasmin system (represented in Figure 2). In milk, it occurs mainly as the inactive precursor plasminogen, with bulk raw milk containing 0.07-0.15 μg mL^{-1} plasmin and 0.7-2.4 μg mL^{-1} plasminogen (Rollema et al. 1981). Plasmin is classified as a serine protease, carrying the amino acid serine at the active site (Grufferty and Fox 1988). It is quite heat-stable, and is known to survive pasteurization processes (Metwalli et al. 1998) and even UHT-treatment (Alichanidis et al. 1986). Nevertheless, it is less heat resistant than *Pseudomonas* proteases that retain 73% of their activity when conditions are applied that completely destroy plasmin_(Marchand et al. 2008).

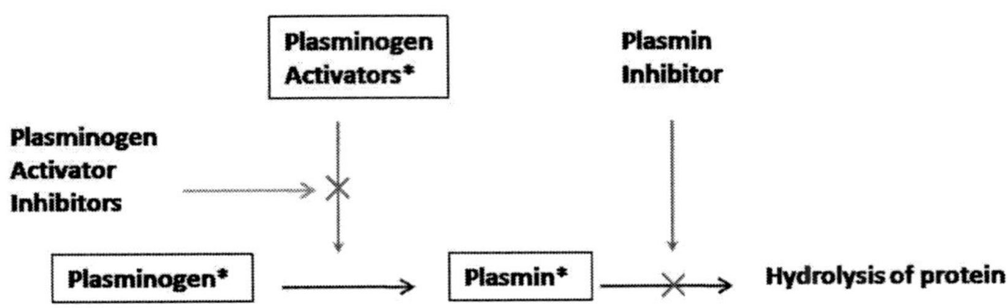

Figure 2. Plasmin system (based on Prado et al. 2006). *: heat-stable, ☐: associated with the casein micelle.

Plasmin hydrolyses α_{S2}- and β-caseins and to a lesser extent α_{S1}-caseins, but has little or no activity on the whey proteins β-lactoglobulin and α-lactalbumin. Reports on the hydrolysis of κ-casein, however, are conflicting (Datta and Deeth 2001). There is also conflicting evidence about the role of plasmin in gelation upon storage of UHT milk (Datta and Deeth 2001) and the importance of plasmin in cheese ripening is still under debate. The latter probably depends on the cheese variety, being somewhat more important in the ripening of high-pH cheese (*e.g.,* Camembert) than in low-pH cheese (*e.g.,* Mozarella) (Bastian and Brown 1996).

While plasmin is the principal indigenous protease in good-quality milk, increasing evidence is now becoming apparent that other proteases including cathepsins and elastase, are also active, especially in milk with a high somatic cell count. Their effect on the quality of milk products however, has been far less intensively studied (Kelly et al. 2006). Cathepsin D appears to be able to at least partially survive commercial pasteurization processes. Furthermore, increasing evidence for a role for this enzyme in proteolysis during cheese ripening is becoming apparent (Hurley et al. 2000).

The enzymes of psychrotolerant bacteria are probably more active and significant during cold storage of milk than indigenous enzymes like plasmin, that may then lose activity due to autolysis (Crudden et al. 2005;Guinot-Thomas et al. 1995). The proteases produced by many psychrotolerant microorganisms are usually extracellular endopeptidases that can be classified as metalloproteinases (Cousin 1989). With only rare exception, the proteases isolated from psychrotolerant microorganisms can be classified either as alkaline or neutral metalloproteases, with a specificity for large hydrophobic amino acid residues (Morihara 1974).

It is generally agreed that whey proteins are not degraded by proteases produced by psychrotolerant microorganisms in raw milk. There are some reports of minor whey protein degradation, but never to the extent as for caseins and it usually takes more time to occur. Their specific secondary and tertiary structure and globular nature probably make it difficult for microbial proteases to degrade them (Cousin 1989).

Fat Content

Aside from milk proteins, milkfat represents another important fraction in milk. Milk is an emulsion or colloid of butterfat globules within a water-based fluid (the milk serum or whey). The major lipid components in milk are triacylglycerols (triglycerids) (98%), but additionally there are small amounts of diglycerids, monoglycerids, cholesterol ester, cholesterol, free fatty acids (FFA) and phospholipids (Cousins and Bramley 1981;Mabbit 1981). More than 95% of the milkfat is globular, with each fat globule being surrounded by a membrane consisting of phospholipids and proteins.

The fatty acids of butterfat typically contain 4-18 carbon atoms. Saturated fatty acids account for 75 % of the total fatty acids in bovine milk, with the long-chain fatty acids myristic (C14), palmitic (C16) and stearic (C18) acid being predominant. A further 21% occurs as mono-unsaturated fatty acids of which the most prevalent is oleic acid (C18:1). Most unsaturated fatty acids in raw milk occur in the *cis* conformation. A Swedish study shows a presence of approximately 2.7% *trans* fatty acids (such as vaccenic acid (C18:1 t11) and rumenic acid (C18:2 c9t11)) in raw milk (Månsson 2008). Even though *trans* fatty acids are considered to be a possible health risk (with respect to cardiovascular disease, inflammation, body weight, insulin sensitivity and even cancer), public health implications of consuming ruminant *trans* fatty acids are thought to be relatively limited (as reviewed by Mozaffarian et al. 2009).

Only 4 g/100 g of the milkfatty acids are polyunsaturated, occurring mainly as linoleic (C18:2) and linolenic (C18:3) acids, though variations can occur according to the cow's diet (Mansbridge and Blake 1997).

Although FFA due to lipolysis of milkfat, are important for the development of cheese flavour, excessive lipolysis resulting from heat resistant bacterial lipases, can cause rancid off-flavours in cheeses with a long shelf life, possibly already after a period of ripening of 2 to 3 months (Cousin 1982). Also, lipolysis is linked to some technological consequences in cheese production: the released FFA (and mono- and diglycerids) are known to inhibit starter bacteria such as *Streptococcus lactis* and *Streptococcus cremoris*, thus retarding acidification (Deeth and Fitz-Gerald 1983).

FFA that are formed due to the action of lipases, particularly those of short and medium chain length (C4-C12), have strong flavours, which are mostly considered undesirable (Scanlan et al. 1965). Several terms have been used to describe these lipolytic and oxidized flavour defects such as 'rancid', 'bitter', 'goaty', 'soapy', 'unclean' and 'butyric' (Shipe et al. 1978). Even-numbered fatty acids (C4, C6, C8, C10 and C12) are the major flavour contributors and long-chain fatty acids C14 and C18 contribute little, if any, flavour (Scanlan et al. 1965) as do very short chain FFA (Cl, C2 and C3) (Kolar and Mickle 1963). Although none of the FFA seem to have a predominant flavour effect, some organoleptic variations may occur, *e.g.,* 'rancid', 'butyric' and 'goaty' flavours are principally caused by C4 and C6 FFA whereas C10 and C12 FFA are mostly responsible for 'soapy' and 'bitter' flavours of lipolysed milk and butter (Deeth and Fitz-Gerald 1983;Woo and Lindsay 2006). Furthermore, unsaturated FFA are subject to oxidation, resulting relatively quickly in a rancid flavour. The flavour of whole milk with an elevated FFA level (>1.5 meq/100g fat) is unacceptable to most people (IDF 1987). An overview of the levels of short- and medium-chain fatty acids (C4–C12) in different types of milk samples and the threshold values are listed in Table 1.

Table 1. Levels of short- and medium chain fatty acids in various milk samples and typical threshold flavour levels (adapted from Chen et al. 2003)

FFA	Concentrations (µmol mL^{-1}) found in			Flavour treshold in milk (µmol mL^{-1})
	Pasteurized milk	UHT milk	Rancid milk	
$C_{4,0}$	0.02	0.15	0.31-0.97	0.28
$C_{6,0}$	0.01	0.05	0.14-0.42	0.12
$C_{8,0}$	0.01	0.03	0.06-0.19	0.05
$C_{10,0}$	0.02	0.04	0.16-0.46	0.04
$C_{12:0}$	0.02	0.03	0.13-0.32	0.04

Lipolytic spoilage of heat treated milk is expected only in products which are stored for a rather long period and in which the fat is susceptible to lipolysis, such as UHT milk (Mottar et al. 1979). Cream and butter have a high lipolytic spoilage potential due to their high fat content and the preference of psychrotolerant lipases to act on the cream phase of milk (Stead 1986). Butter can become rancid because of growth of lipolytic bacteria due to a bad distribution of moisture (Deeth and Fitz-Gerald 1983) or due to residual heat resistant lipolytic activity after pasteurization (Nahsif and Nelson 1953). And although no bacterial growth is possible at a water activity (a_w) below 0.9, powdered milk products with an a_w as low as 0.6 and derivatives can still be spoiled due to the action of hearesistant bacterial lipase (Shamsuzzaman et al. 1989). Lipases are produced concomitantly with proteases by the same bacterium and are generally regarded to be more heat-stable than the proteases (Chen et al. 2003). However, recent data do not seem to confirm this dogma (unpublished results, De Jonghe et al.)

Lipolytic enzymes is a description for groups of enzymes, including esterases (or carboxylases), true lipases (or triacylglycerol acylhydrolases) and phospholipases.

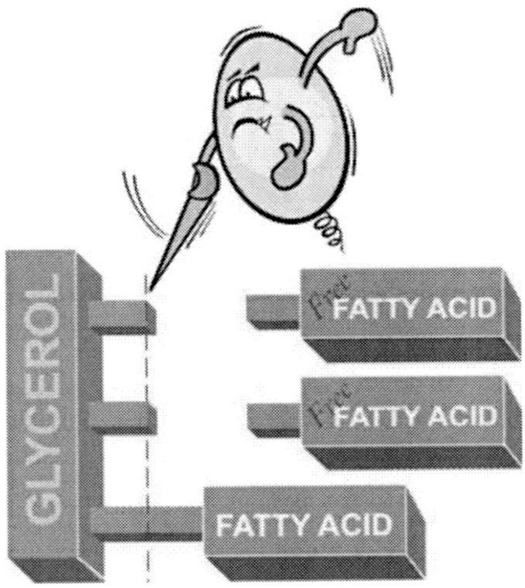

Figure 3. Enzymatic reaction of a lipolytic enzyme catalyzing hydrolysis of a triacylglycerol substrate. *Source: Dairy Processing Handbook* (Bylund, 1995).

Lipases are enzymes that catalyse the hydrolysis of carboxyl ester bonds present in triglycerids (triacylglycerols), the major lipid component of milk. The products of this so called 'lipolysis' are free, non-esterified fatty acids, mono- and diglycerids and in some cases even glycerol (Figure 3).

Lipases act at the lipid-water interface of emulsions of long-chain (≥10), insoluble triglycerids while the related esterases act on esters of short chain fatty acids and soluble esters, although lipases may also hydrolyse such substrates (Jaeger et al. 1994).

The glycoprotein lipoprotein lipase (LPL) accounts for most of the indigenous lipolytic activity in fresh bovine milk, which contains LPL levels varying between 0.5 and 2.0 mg L^{-1} (Chen et al. 2003;Olivecrona 1980;Olivecrona et al. 2003). This enzyme is mainly associated with the casein micelle through electrostatic (Olivecrona et al. 2003) and hydrophobic interactions (Fox et al. 1967). It shows positional specificity, preferably hydrolyzing fatty acids from the 1- and 3-positions of the triglyceride molecule, but no fatty acid specificity. Because short chain fatty acids are concentrated at the 3-position of bovine milk triglycerids, it appears as if LPL shows a general preference towards triglycerids containing short chain fatty acids (Deeth 2006). This is also reflected by a higher indigenous lipolytic activity in milk from thrice daily milking and automatic milking equipments, since a higher milking frequency leads to an increased de novo synthesis of short chain fatty acids (Abeni et al. 2005;Klei et al. 1997;Slaghuis et al. 2004).

The effects of LPL are mostly associated with fresh milk and cream, the effects in cheese and butter being obvious at manufacture. In addition, the heat-labile milk LPL is destroyed upon HTST pasteurization or more severe heat treatments, thus limiting the importance of this enzyme in spoilage of dairy products (Farkye et al. 1995).

Even though the total LPL activity in raw milk is sufficient to cause rapid hydrolysis of a large proportion of the fat, this does not happen in reality, since the lipase cannot readily access the fat which is encapsulated by a phospholipid membrane, called the milkfat globule membrane (MFGM). Lipolysis can be categorized into two types: spontaneous and induced lipolysis. Spontaneous lipolysis is defined by the FFA level in untreated milk (except for cooling) immediately after milking (Deeth and Fitz-Gerald 1983). It occurs at the farm only, with milk of some individual cows being more susceptible than others. The biochemical basis of spontaneous lipolysis remains poorly understood. Furthermore, milk susceptible to spontaneous lipolysis is also more susceptible for induced lipolysis, which is initiated by cold mechanical disruption of the MFGM so that the enzyme can now easily access the fat fraction of the milk. This can happen either mechanically, due to agitation, pumping, stirring and freezing/thawing of milk or by enzymatic means, such as by phospholipases or glycosidases (Figure 4).

Homogenization of milk is a process by which fat globules in fluid milk are broken into sizes small enough (1-8 μm in raw milk to 0.3-0.8 μm in homogenized milk) not to rise in the milk so that cream cannot be formed under normal milk storage conditions. Because the smaller fat globules are now surrounded by a protein coat, this could help the fat and milk proteins to partially regain a protective interface (Mabbit 1981). However, it seems to be of minor importance to LPL, since homogenization takes place immediately before or after the heating process by which LPL is inactivated (Deeth 2006).

In the dairy industry, not all undesirable lipolysis is caused by LPL. Some important lipase-producing bacterial genera include *Bacillus*, *Pseudomonas* and *Burkholderia* (Gupta et al. 2004).

Bacterial lipases are serine hydrolases that share a similar folding pattern (called the α/β-hydrolase fold) and have a common structural motif, namely a highly conserved pentapeptide consensus motif (G-X-S-X-G) within the catalytic triade that consists of two conserved glycines and a conserved serine, aspartate or glutamate and a histidine residue (Derewenda and Derewenda 1991;Gupta et al. 2004).

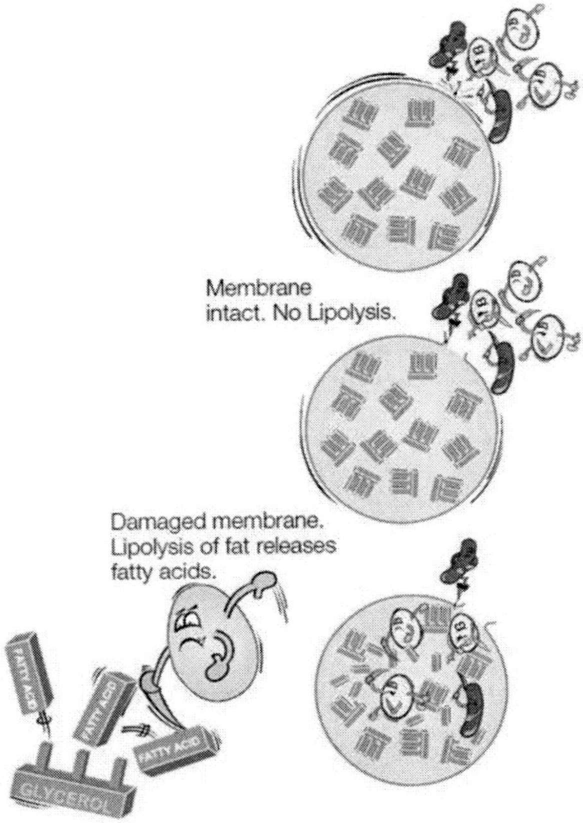

Figure 4. Correlation between phospholipolytic (in red) and lipolytic (in yellow) activity. *Source: Dairy Processing Handbook* (Bylund, 1995).

In general, they have molecular masses ranging from 30 to 50 kDa (with an exception for certain *Bacillus* lipases belonging to family I.4 (Table 2) with a molecular mass of approximately 20 kDa) and pH optima between 7 and 9 (Chen et al. 2003). Most of them have specificity for the 1- and 3-positions of triacylglycerols, and some hydrolyse diacylglycerols and monoacylglycerols faster than triacylglycerols (Macrae 1983).

Bacterial lipases and esterases are grouped into eight different families based on amino acid sequence homology and some fundamental biological properties. The largest family comprises the bacterial true lipases (family I; Table 2) that were formerly ordered in so-called *Pseudomonas* groups 1, 2 and 3 since *Pseudomonas* lipases were probably the first to be studied.

Table 2. The family of true lipases (family I). Amino acid sequence similarities were determined with the program MEGALIGN (DNASTAR), with the first member of each family (subfamily) arbitrary set at 100%. (adapted from Jaeger and Eggert 2002)

Family	Subfamily	Enzyme-producing strain	Accession no.	Similarity (%)	
				Family	Subfamily
I	1	*Pseudomonas aeruginosa* (LipA)	D50587	100	
		Vibrio cholerae	X16945	57	
		Pseudomonas aeruginosa (LipC)	U75975	51	
		Acinetobacter calcoaceticus	X80800	43	
		Pseudomonas fragi	X14033	40	
		Pseudomonas wisconsinensis	U88907	39	
		Proteus vulgaris	U33845	38	
	2	*Burkholderia glumae*	X70354	35	100
		Chromobacterium viscosum	Q05489	35	100
		Burkholderia cepacia	M58494	33	78
		Pseudomonas luteola	AF050153	33	77
	3	*Pseudomonas fluorescens SIK W1*	D11455	14	100
		Serratia marcescens	D13253	15	51
	4	*Bacillus subtilis* (LipA)	M74010	16	100
		Bacillus pumilus	A34992	13	80
		Bacillus licheniformis	U35855	13	80
		Bacillus subtilis (LipB)	C69652	17	74
	5	*Geobacillus stearothermophilus* L1	U78785	15	100
		Geobacillus stearothermophilus P1	AF237623	15	94
		Geobacillus thermocatenulatus	X95309	14	94
		Geobacillus thermoleovorans	AF134840	14	92
	6	*Staphylococcus aureus*	M12715	14	100
		Staphylococcus haemolyticus	AF096928	15	45
		Staphylococcus epidermidis	AF090142	13	44
		Staphylococcus hyicus	X02844	15	36
		Staphylococcus xylosus	AF208229	14	36
		Staphylococcus warneri	AF208033	12	36
	7	*Propionibacterium acnes*	X99255	14	100
		Streptomyces cinnamoneus	U80063	14	50

Table 3. Comparison of the characteristics of milk lipoprotein lipase (LPL) and lipases from psychrotolerant bacteria (adapted from Deeth 2006)

Milk LPL	Lipases from psychrotolerant bacteria
Destroyed by HTST pasteurization	Stable to HTST and even to UHT-treatment
MFGM acts as a barrier to lipid substrate	MFGM presents no barrier
Effect mostly associated with fresh milk and cream	Effect mostly associated with stored products – UHT milk, cheese, butter, milk powders
Effect in cheese/butter obvious at manufacture	Effect in cheese/butter obvious only after storage

Because of taxonomic revisions and the description of many lipases from other genera, a revised classification of true lipases was proposed by Arpigny and Jaeger (1999) and updated by Jaeger and Eggert in 2002 (Table 2). Bacterial lipases have different characteristics from LPL as summarized in Table 3.

Apart from the difference in heat stability, the most striking difference is that bacterial lipases in reality appear not to be hindered by the MFGM. The mode of access and mechanism of this activity are not yet known (Deeth and Fitz-Gerald 1994), but a possible explanation is the action of accompanying enzymes such as phospholipases (Mabbit, 1981) as demonstrated in Figure 4Fout! Verwijzingsbron niet gevonden.. Phospholipases, especially type C or lecithinase which hydrolyses phosphatidylcholine in the MFGM, are produced by many types of bacteria including *Pseudomonas*, *Bacillus* and *Clostridium* (Cousin 1989).

These extracellular phospholipases are able to withstand various heat treatments (even UHT-treatment) of milk (Deeth and Fitz-Gerald 1983;Griffiths 1983;Koka and Weimer 2001).

Few reports exist on the degradation of the MFGM due to the activity of bacterial glycosidic enzymes that can remove the sugar residues from the outer layer of the MFGM, making the underlying proteins, lipids and phospholipids more accessible to other hydrolytic enzymes such as proteases, lipases and phospholipases, respectively (Marin et al. 1984).

Sugar Content

Milk sugar, the disaccharide lactose (β-D-galactopyranosyl-(1-4)-D-glucopyranose), is the predominant carbohydrate of milk. In addition, very low concentrations of monosaccharides (a.o. glucose and galactose), oligosaccharides and protein-bound carbohydrates (*e.g.* in κ-casein) can be present (Banks et al. 1981).

At room temperature, milk undergoes natural souring caused by lactic acid produced from fermentation of lactose by fermentative lactic acid bacteria (LAB) (Mabbit 1981).

This accumulation of acid decreases the pH of the milk and causes the casein to coagulate and curdle into curds (i.e. large, white clumps of casein and other proteins) and whey. This phenomenon is used for processing of many milk products such as yoghurt and cheese: lactose is enzymatically degraded into its sugar building blocks (galactose and glucose) by the enzyme β-galactosidase. There are two main fermentation pathways that are used to classify LAB genera: homolactic LAB (*e.g., Lactococcus, Enterococcus, Streptococcus*, *Pediococcus* and group I lactobacilli) catabolize one mole of glucose in the Embden-Meyerhof-Parnas (EMP) pathway to ultimately yield two moles of lactic acid, whereas heterofermentative LAB (*e.g,. Leuconostoc, Oenococcus, Weissella* and group III lactobacilli) mainly use the phosphoketolase pathway resulting in the production of one molecule of carbon dioxide, one molecule of ethanol, and one molecule of lactic acid as represented in Figure 5.

The phenomenon of lactose fermentation is used to our advantage in making many milk products such as yoghurt and cheese through addition of starter cultures (generally LAB), since the natural microbiota of milk is either inefficient and uncontrollable or is destroyed altogether by the heat treatments (pasteurization, thermisation) given to the milk. Starter cultures can be divided into two groups: primary and secondary microbiota. Products undergoing fermentation by only primary microbiota are called 'unripened' milk products (e.g. unripened cheeses such as cottage cheese, cream cheese, Mozarella and quark) and those processed by both primary and secondary microbiota are called 'ripened' milk products (e.g. soft and hard ripened cheeses). Primary microbiota are fermentative LAB which cause the milk to curdle. Secondary microbiota include several different types of bacteria (*Lactococcus lactis* and *Leuconostoc cremoris* are used most often) to produce various cheeses.

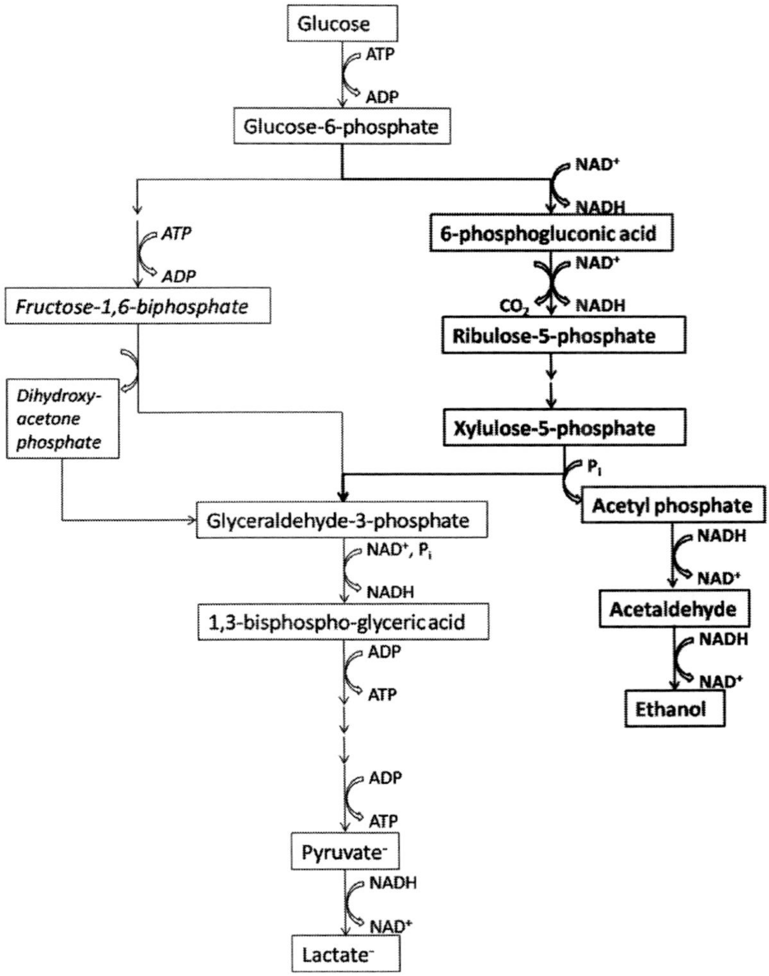

Figure 5. The fermentation of glucose in homofermentative (italics) and heterofermentative (bold) lactic acid bacteria. Shared pathways for homo- and heterofermentative fermentation are indicated in regular font.

Aside from uncontrolled and therefore undesired growth and lactose fermentation by LAB, the psychrotolerant microbiota of refrigerated raw milk also contains fermentative bacteria, *i.e.,* facultative anaerobic organisms such as certain *Bacillus* species, that can cause undesirable lactose fermentation with concomitant acid and off-flavour development. However, it seems more likely that psychrotolerant organisms contribute rather in an indirect way through enzymatic activity which can have both stimulating and inhibiting effects on starter cultures as the latter may benefit from a greater accessibility of nitrogen sources through proteolytic activity or, contrarily, be inhibited by FFA (and partial glycerids) released upon lipolytic activity (Deeth and Fitz-Gerald 1983;Fox 1981;Mabbit 1981).

How Do the Spoilage-Causing Bacteria Get into the Milk?

The initial microbiota of raw milk (*i.e.,* the microbiota that is present immediately after milking) can vary in numbers between $<10^3$ to $>10^6$ cells per mL (Cousins and Bramley 1981) and in diversity as influenced by the amount of hygienic measures that are taken into account during the various stages of milk handling (Verdier-Metz et al. 2009). During storage and transport throughout the dairy chain, there is a possible outgrowth of the microbiota already present in raw milk. Since the adoption of refrigerated bulk tanks for the collection and storage of raw milk prior to processing to prevent outgrowth of LAB and pathogens, the predominant organisms in raw milk are now psychrotolerant bacteria, of which the majority is destroyed by pasteurization, but not their produced extracellular enzymes that withstand various heat treatments (Cogan 1977).

Entry at the Farm Level

At the farm level, there are three main sources of microbial contamination in milk: from within the udder, the exterior of the teats and udder and the milking and storage equipment (Cousins and Bramley 1981). Raw milk from cows suffering from mastitis is more susceptible to contamination since the bacteria responsible for this udder infection can multiply inside the udder, thus infecting the glandular tissue. Insufficient cleaning of the teats before milking can contaminate the raw milk with bacteria that are present in soil, faeces, straw etc. with which the teats are fouled. Psychrotolerant bacteria, both potentially pathogenic bacteria as well as bacteria that can interfere with processing of the milk, have soil, water, animal and plant material as natural habitat (Cousin 1982).

Plant materials that are commonly used for animal feed (*e.g.,* grass, hay) may contain over 10^8 psychrotolerant bacteria per gram (Thomas 1966) and the bedding materials on which cows are housed in the winter show a count of 10^9 psychrotolerant bacteria per gram on average (Cousins and Bramley 1981). The milking equipment, storage tanks and milk tankers are generally considered the major contamination source for psychrotolerant bacteria (Cousin 1982). The equipment is mainly made from stainless steel, glass, plastics and rubber. The use of untreated water supplies for the final rinse of the milking equipment may contribute to contamination of raw milk with psychrotolerant microorganisms (dominated by *Pseudomonas*, *Achromobacter*, *Alcaligenes* and *Flavobacterium* (Thomas 1966)). Because psychrotolerant bacteria isolated from water are proven to be vigorous producers of extracellular enzymes and grow rapidly in refrigerated raw milk, contaminated water can be considered an important source of milk spoilage bacteria regardless of the possible low initial contamination level (Cousin 1982). A likely reservoir from which contamination of these water supplies originate, is the soil (Thomas 1966). Even though proper cleaning of the milking equipment effectively reduces contamination from these sources, the rubber materials used to connect different pipelines are quite susceptible to deterioration caused by a combination of high cleaning temperatures and strongly oxidizing products in the disinfectants (used to kill off a considerable fraction of spores).

The resultant microscopic cracks and cuts form an ideal attachment place for the formation of biofilms (Morse et al. 1968). These multispecies structures (harbouring among others *Bacillus* and *Pseudomonas* species) often possess greater combined stability to mechanical treatments and resilience to chemical sanitizers than do the constructing individual species (Simões et al. 2009).

Besides psychrotolerant microorganisms, various studies have been performed on the contamination sources of aerobic spore-formers in raw milk. Most research focuses on *Bacillus cereus*, that is considered to be the most important spoilage organism in the dairy industry (as discussed in section 3.2.2). Different studies point to different contamination sources in the milking environment responsible for the entry of *B. cereus* cells or spores in raw milk as shown in Table 4.

The general consensus as major contamination source for aerobe spore-formers now appears to be soil (particularly in the grazing season) and feed, supplemented with occasional contaminations *e.g.,* from the milking equipment and silos at the dairy plant (Svensson et al. 2004).

Table 4. Contamination sources of *Bacillus cereus.* [a] during wet summers, [b] during winter period, [c] could not be excluded, particularly during summer, **Bacillus* species in general

Reference	Source
Billing and Cuthbert (1958)	soil[a], hay[b], dust[b]
Labots and Hup (1964)	soil, feed, faeces, milking equipment[c]
Davies and Wilkinson (1973)*	udder hygiene, soil, bedding material
Stewart (1975)	feed, bedding material, dust
Waes (1976)*	udder hygiene
Barkley and Delaney (1980)	teat cups contaminated with spent barley grain from the brewing industry
Palmer (1981)	air
Stadhouders and Jørgensen (1990)*	udder hygiene (combined with construction of milking machine)
te Giffel et al. (1995)	soil, faeces
Christiansson et al. (1999)	soil
Lukasova et al. (2001)*	feed, udder hygiene
Magnusson et al. (2007)	feed via faeces
Vissers et al. (2007)	feed via faeces

Outgrowth throughout the Cold Dairy Chain

Currently, there are no general official standards for spores and psychrotolerant bacteria in raw milk in the EU. In the Netherlands, a spore count of 10^3 spores from butyric acid bacteria (BAB) per liter (in order to get a concentration of less than 10^1 BAB spores per liter after bactofugation) is considered a good criterion for good quality raw milk.

For psychrotolerant bacteria, unexplained problems in milk processing can frequently be attributed to changes in the ratio of psychrotolerant versus total bacterial count which is normally approximately 16.7% for bulk tank milk (Cempirkova 2002).

Søgaard and Lund (1981) described that the number of psychrotolerant versus total bacteria increased from 4.1% on the farm to 6.2% on the milk tanker to 13.9% in the dairy bulk tank in winter time and correspondingly from 16.7, 21.9 and 78.1% in the summer period, with a final count for psychrotolerant microorganisms in the dairy bulk tank 5.8×10^3 and 9.6×10^4 CFU (colony forming units) per mL milk for winter and summer, respectively.

This seasonal difference could be attributed to a fivefold higher initial contamination in the farm bulk tank in the summer (Søgaard and Lund 1981), confirming that a high initial contamination results in a rapid outgrowth of psychrotolerant bacteria in raw milk because more bacteria are actively growing (Thomas 1966).

Four factors are important in the pursuit for a better microbiological quality of the raw milk throughout the dairy chain: (i) the amount of bacteria that are present in the raw milk, (ii) the nature of bacteria, (iii) the storage temperature and (iv) the storage time.

Hygiene in all aspects of milk handling, strict maintenance of refrigeration at 4°C or lower, minimization of the storage period of raw milk, combined with a suitable method to remove or kill as many microorganisms as possible and followed up by an effective HACCP system, are therefore important parameters of primary concern in the dairy industry.

Good hygienic practices can lead to a decrease in the amount of (harmful) bacteria present in raw milk.

Nonetheless, a study performed by Richard (1981) showed that intensive washing of milking equipment and udder preparation result in raw milks containing a high load of spoilage microorganisms such as *Pseudomonas* spp. and coliforms.

The use of a lower storage temperature has led to believe that the milk could be stored for a longer period before processing at the dairy factory. However, this prolonged cold storage of raw milk prior to processing creates a selective advantage for psychrotolerant populations, that can grow out after a storage time of less than 24 h at 4°C (Lafarge et al. 2004).

Moreover, this lower storage temperature is not consistently reached. A recent study by De Jonghe et al. (2011) clearly demonstrates the importance of low storage temperatures throughout the cold dairy chain on both total colony count and *Pseudomonas* count, the latter being the predominant psychrotolerant micro-organism in raw milk (Adams et al. 1975;Garcia et al. 1989;Sørhaug and Stepaniak 1997). This particular study reports a possible surplus of 2 log CFU per mL raw milk in both total colony count and *Pseudomonas* count at the end of cold storage prior to processing when milk is stored suboptimally (Figure 6). Furthermore, a low total colony count does not necessarily guarantee a low total spore count as demonstrated by Rombaut et al. (2002).

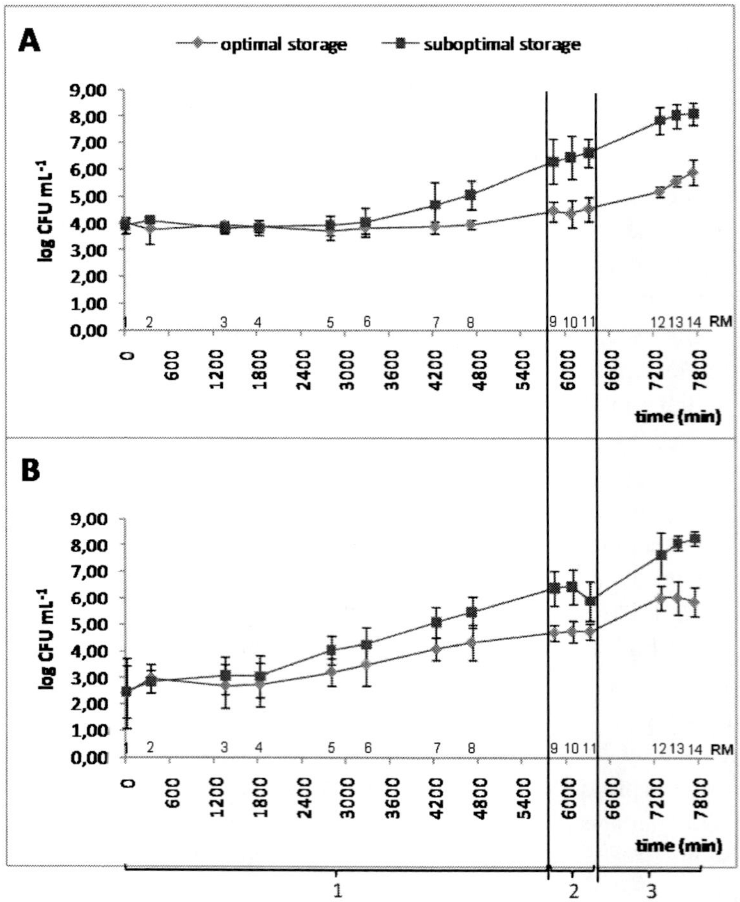

Figure 6. Total colony count (A) and total *Pseudomonas* count (B) as determined upon simulation of the cold dairy chain. 1: simulation of storage in the farm tank, 2: simulation of storage during transport, 3: simulation of storage at the dairy plant (De Jonghe et al. 2011) (Copyright © American Society for Microbiology, Applied and Envirmonmental Microbiology, 2011, 77:460-470, doi:10.1128/AEM.00521-10).

It's Not over Yet: Effect of Postprocessing

Processing of the raw milk does not effectively kill all microorganisms (except for sterilization): bacterial spores cannot be destroyed by conventional heating processes, such as pasteurization (Andersson et al. 1995). On the contrary, bacterial load may even be higher when treatments are applied that are more severe than those required for pasteurization, as several studies indicate that higher pasteurization temperatures result in higher bacterial numbers (belonging to the genera *Bacillus* and *Paenibacillus*) in fluid milk products (Hanson et al. 2005;Ranieri et al. 2009) as spores are stimulated to germinate upon these heating conditions. Being the most intensely studied aerobe spore-forming organism in milk, spores from *B. cereus* are well-known for surviving different pasteurization conditions (Aires et al. 2009;Novak et al. 2005). Spores of some species are even known to survive UHT-treatment (Scheldeman et al. 2006). The most heat resistant species are *Geobacillus stearothermophilus*, *Bacillus sporothermodurans* and *Paenibacillus lactis* (Muir 1989;Pettersson et al. 1996;Scheldeman et al. 2004;2005;2006).

Furthermore, after processing milk can become recontaminated with microorganisms when exposed to contaminated air, mainly during the filling step (Eneroth et al. 1998). This phenomenon is known as post-pasteurization contamination (PPC) (Schröder 1984). Since pasteurization affects the growth rate of spoilage microbiota by destroying the inhibitor mechanisms that are naturally present in milk (the lactoperoxidase system, among others) (Wolfson and Sumner 1993), post-pasteurization contaminants may be able to grow more rapidly in pasteurized milk than in the raw product. For pasteurised and Extended Shelf Life (ESL) milk, the filling machine has been shown as the main source of recontamination (Rysstad and Kolstad 2006). Also the bulk milk tanks where the pasteurized milk is stored until filling can be held responsible for sporadic outbreaks of relatively high contamination (Schröder 1984). PPC can be substantially reduced or even eliminated through aseptic filling that uses pre-sterilised containers that are then filled with cold product in a cold environment in commercially sterile conditions, followed by closure in a totally sterile environment (Stepaniak 1991).

Which Micro-Organisms to Fear?

Psychrotolerant bacteria are defined as bacteria that are able to grow at 7°C or less, regardless of their optimal growth temperature (Suhren 1989). They have become an escalating problem in the dairy industry ever since the introduction of refrigerated storage throughout the dairy chain, because of their selective advantage over non-psychrotolerant bacteria. Both Gram-negative and Gram-positive psychrotolerant bacteria are implicated in milk spoilage through the production of spoilage enzymes such as lipases and proteases (Sørhaug and Stepaniak 1997).

Gram-Negative Spoilage Organisms

In milk produced under sanitary conditions, the typical bacteria of the udder surface, mainly *Micrococcaceae*, predominate and less than 10% of the total microbiota are psychrotolerant microorganisms, but this percentage can mount up to 75%-90% under unsanitary conditions (Adams et al. 1975;Kurzweil and Busse 1973;Thomas and Thomas 1973). The main psychrotolerant aerobic bacteria which contaminate raw and pasteurized milk are primarily aerobic Gram-negative rods belonging to the *Pseudomonaceae* with approximately 65-70% of psychrotolerant isolates from raw milk assigned to the genus *Pseudomonas* (Garcia et al. 1989). Other genera present include *Aeromonas*, *Acinetobacter*, *Alcaligenes*, *Chromobacterium*, *Flavobacterium* and *Serratia* (Champagne et al. 1994;Cousin 1982;Lafarge et al. 2004). Under the low temperature conditions throughout the dairy chain, members of the genus *Pseudomonas* are able to grow out and dominate the microbiota found in raw milk (Sørhaug and Stepaniak 1997). This may be explained because *Pseudomonas* members show the shortest generation times at 0-7°C (Chandler and McMeekin 1985). Furthermore, *Pseudomonas* spp. are able to colonize the processing line by adhering strongly to the surface of the milk processing equipment. This may enable them to persist unless removed by proper cleaning and sanitizing procedures (Bishop and White 1986;Cousin 1982). In the summer season, there is an increase in total psychrotolerant count, but no typical seasonal pattern was observed in the incidence of *Pseudomonas* (Garcia et al. 1989). However, a seasonal pattern in the proteolytic capacity of *Pseudomonas* isolates from raw milk was demonstrated by Marchand et al. (2009a).

P. fluorescens has traditionally been accepted as the most important spoilage organism (Dogan and Boor 2003;Jayarao and Wang 1999). Nowadays, the importance of *P. fluorescens* in milk spoilage is under debate as it seems to be overestimated in the past due to an incorrect identification (Marchand et al. 2009a).

Marchand et al. (2009a) identified *Pseudomonas lundensis* and *Pseudomonas fragi* members as the most important proteolytic spoilers in raw milk based on a thorough identification of the strains using a polyphasic approach. A recent study by De Jonghe et al. (2011) acknowledged the predominant presence and spoilage capacity of *P. fluorescens*-like and *P. gessardii*-like organisms, being closely related but clearly distinct from the *P. fluorescens* type strain.

As pseudomonads are well-known spoilage organisms, a lot is documented about their spoilage enzymes. Though optimal enzyme synthesis occurs in the majority of psychrotrotolerant bacteria at 20-30°C, considerable synthesis occurs even at lower temperature, for example, production of extracellular protease by *Pseudomonas fluorescens* at 5°C was 55% of that produced at 20°C (McKellar 1982). Furthermore, the enzymes remain active at temperatures well under their optimum temperature, for instance even at 2°C for *P. fluorescens* (Braun et al. 1999).

In contrast to lipolytic enzymes, the majority of *Pseudomonas* species produce only one heat resistant type of protease that is thought to be responsible for the spoilage of milk (Dufour et al. 2008;Fairbairn and Law 1986;Marchand et al. 2009b): the alkaline metalloprotease AprX protease that is widespread throughout the genus *Pseudomonas* (Chabeaud et al. 2001;Kumeta et al. 1999;Liao and McCallus 1998;Marchand et al. 2009b). It has a molecular mass of approximately 45 kDa (Dufour et al. 2008;Koka and Weimer 2001;Marchand et al. 2009b) and it belongs to the highly conserved serralysin family that is characterized by a zinc binding motif, a calcium binding domain containing four glycine rich repeats (G-G-X-G-X-D), a high content of hydrophobic amino acids and no cysteine residues (Kumeta et al. 1999;Rawlings and Barrett 1995). It is encoded by the *aprX* gene which lies on the *aprX-lipA* operon as demonstrated for *P. fluorescens* strain B52 (McCarthy et al. 2004;Woods et al. 2001).

Even though *Pseudomonas* species are easily inactivated by various heat treatments, an important fraction of the spoilage enzymes that they produce during growth, remains active because of their resistance to high temperatures. *Pseudomonas* species are known to produce heat-stable spoilage enzymes that retain significant activity even after UHT processing and production of milk powders (Chen et al. 2003). These enzymes can then cause spoilage and structural defects in pasteurized and UHT-treated milk and milk-powder derived products (chocolate, deserts etc.).

Thermostability of *P. fluorescens* proteases is the most intensively studied but strains belonging to other *Pseudomonas* species have also been proven to retain approximately 10% of their original activity after exposure to 140°C for 5 s (Kroll 1989). A recent study by Marchand et al. (2009a) showed *P. fragi* and *P. lundensis* as the most important producers of heat-stabile proteases.

Even though the occurrence of heat-stable lipases is much less extensively studied than the occurrence of heat-stable proteases, heat-resistance is believed to be a common characteristic of lipases from psychrotolerant microorganisms (Andersson et al. 1979;Cogan 1977;Cousin 1982;Shelley et al. 1986;Shelley et al. 1987). Griffiths et al. (1981) found residual lipase activity of over 10% (mean value 30%) after exposure to 140°C for 5 s in strains belonging to a wide variety of *Pseudomonas* species (*P. fluorescens*, *P. stutzeri*, *P. putida* and *P. fragi*) (Griffiths et al. 1981). However, recent data do not support this overall heat-stability of *Pseudomonas* lipases (unpublished data, De Jonghe et al.).

A unique feature of both proteases and lipases of psychrotolerant *Pseudomonas* species, is their sensitivity toward low temperature inactivation (LTI) (Kroll 1989), meaning that they are rapidly irreversibly inactivated just above the optimum temperature for activity.

For proteases, the formation of enzyme-casein aggregates is proposed as an explanation for this phenomenon rather than an autolytic mechanism due to unfolding of the protein chain into a more sensitive conformation (Chen et al. 2003). For lipases however, the mechanism for LTI still remains unclear (Kroll 1989;Sørhaug and Stepaniak 1997): hydrolysis by proteinases or inactivation by aggregation with caseins has been suggested (Gasincova et al. 1994). It seems that lipases are more sensitive to this type of inactivation than proteases (Griffiths et al. 1981).

Thermoduric Spoilage Organisms

In the USA, an estimated 25% of all shelf life problems in conventionally pasteurized milk and cream products is linked to thermoduric psychrotolerant organisms (Meer et al. 1991), among which aerobic spore-formers belonging to the genus *Bacillus* and relatives (*i.e., Bacillus sensu lato* (*s.l.*)) dominate other psychrotolerant bacteria such as *Arthrobacter*, *Alcaligenes*, *Microbacterium*, *Micrococcus*, *Streptococcus*, *Corynebacterium* and *Clostridium* (Hayes and Boor 2001;Sørhaug and Stepaniak 1997). *Bacillus* spp. and *Paenibacillus* spp. are of particular concern due to their psychrotolerant properties and spoilage capacity (Coorevits et al. 2008;De Jonghe et al. 2010). The generation times and lag phases of psychrotolerant bacilli at 2-7°C are considerably longer than those of *Pseudomonas* spp. (Chandler and McMeekin 1985), but nonetheless, they can become the dominant microbiota in spoiled pasteurized milk that is stored at 10°C (Meer et al. 1991;Stepaniak 1991). Furthermore their spores cannot be destroyed by conventional processing conditions such as pasteurization and for some species even UHT: *B. cereus* has a $D_{72°C}$-value of 33.5 s as determined in whole milk, which enables it to survive HTST pasteurization (Xu et al. 2006), whereas *B. sporothermodurans* spores from UHT milk isolates are able to withstand even UHT treatment with $D_{140°C}$-values varying between 3.4 and 7.9 s as determined in spiked milk (Huemer et al. 1998).

Spoilage enzymes are produced upon germination of the spores with a maximum synthesis in the late exponential and early stationary phases of growth, before sporulation (Priest 1977). *Bacillus* strains tend to produce both intracellular and extracellular lipases and proteases (both serine and metalloproteases) with comparable thermostability to *Pseudomonas* enzymes, sufficient to withstand any of the heat treatments applied during a milk manufacturing process (Chen et al. 2004;Sørhaug and Stepaniak 1997).

Even though the presence of aerobic spore-forming bacteria can have severe implications for the dairy industry, very little is known about the identity of the most important spoilage causing species, as they are often not further specified (McKellar 1989) or because identification is based on phenotypical and biochemical characteristics of the strains. In this light *Bacillus circulans*, *Bacillus coagulans*, *Brevibacillus laterosporus* (Shehata et al. 1971), *Paenibacillus polymyxa* (Ternström et al. 1993) and *B. cereus* (Overcast and Atmaram 1974) have been implicated in milk spoilage. However the identification of the strains has become outdated and insufficient in view of current taxonomical rearrangements and developments in the aerobic spore-forming microbiota. This was already obvious because 1,6 to 48 % of all aerobic spore-forming isolates obtained from raw milk could not be identified in these studies (Phillips and Griffiths 1986;Sutherland and Murdoch 1994). A recent study by Coorevits et al. (2008) identified the *B. cereus* group, *Paenibacillus polymyxa* and the *B. subtilis* group (more specifically *B. subtilis*, *B. pumilus*, *B. amyloliquefaciens* and *B. licheniformis*) as the predominant aerobic spore-forming spoilers in raw milk, based on a polyphasic identification approach.

Historically speaking, the most important spore-forming spoilage organism in the dairy industry is undoubtedly *B. cereus*, causing defects in pasteurized milk known as 'bitty cream' (floating clumps of fat) due to lecithinase activity and 'sweet curdling' (coagulation of the milk without acidification) due to proteolytic activity (Heyndrickx and Scheldeman 2002). While bacteria other than *B. cereus* and *Bacillus mycoides* produce lecithinase enzymes, only lecithinase-positive *B. cereus* isolates have been shown to produce bitty cream (Owens 1978). Sweet curdling on the other hand, can also be linked to *B. subtilis* and *Br. laterosporus* (Heyndrickx and Scheldeman 2002).

Not much is known about the nature of the proteolytic *Bacillus* enzymes involved in milk spoilage. *Bacillus* species are capable of producing more diverse proteolytic activities than *Pseudomonas* species, and many may produce more than one type of protease (a serine protease and a metalloprotease), the proportions of both enzymes differing among strains (Chen et al. 2004).

In all known lipases from *Bacillus s.l.* (belonging to subfamily I.4 and I.5 as can be derived from Table 2) the first glycine in the pentapeptide consensus motif is replaced by an alanine (A-X-S-X-G instead of G-X-S-X-G as described earlier in this chapter) (Eggert et al. 2000). As a direct consequence to this difference in sequence within the catalytic triade, Eggert et al. (2000) demonstrated a shift in substrate specificity to smaller triglycerids due to steric constraints, which classifies these enzymes into the group of esterases rather than lipases. They also noticed a marked reduction in the thermostability of the enzyme when the alanine was replaced by a glycine (Eggert et al. 2000).

Lipase enzymes from *Bacillus s.l.* are secreted via the Sec machinery (similar to *Pseudomonas* families I.1 and I.2) or by means of the Tat pathway as described for *Bacillus subtilis* LipA (Jaeger and Eggert 2002). As opposed to Gram-negative organisms, no accessory proteins have been described in Gram-positive bacteria, where it seems that N-terminal pro parts of lipases and proteases function as intramolecular foldases that are cleaved off after secretion of the enzyme (Shinde and Inouye 1993).

Spoilage caused by aerobic spore-forming bacteria is not restricted to production of extracellular spoilage enzymes: they are also involved in defective cheese preparation through fermentative growth with gas production as demonstrated recently for *Paenibacillus polymyxa* in Argentinian Cremoso and Mozarella cheeses (Quiberoni et al. 2008). These so-called blowing defects usually arise from growth of mainly *Clostridium tyrobutyricum, Clostridium beijerinckii* and occasionally from *Clostridium sporogenes* and *Clostridium butyricum*. This growth typically leads to ‘late blowing’ defects in semi-hard cheeses, a type of gassy defect that results from the fermentation of lactate to butyric acid, acetic acid, carbon dioxide and hydrogen gas (Klijn et al. 1995;Le Bourhis et al. 2005). It manifests after the cheese has aged for several weeks as opposed to early blowing caused by coliform bacteria (described below). The presence of *C. tyrobutyricum* spores in milk originates from contaminated silage which generally has a high pH that allows growth of clostridia (Dasgupta and Hull 1989).

Another problem associated with fermentative growth of *Bacillus* species is known as “flat sour” defect (acidification without gas production) in evaporated milk, which can result from growth of *Geobacillus stearothermophilus*, *B. licheniformis*, *B. coagulans*, *Paenibacillus macerans* and *B. subtilis* (Kalogridou-Vassiliadou 1992;Speck 1976).

A problem not of immediate product quality, but rather a sterility issue in UHT-milk was observed for the first time in the 1990's. At that time, EC-regulation 92/46 required that the number of colonies counted from incubated (30°C during 15 days) unopened UHT-cartons, should not exceed 10 CFU per 0,1 mL. However, an unknown mesophilic spore-forming microorganism (originally named HRS or HHRS – higly heat resistant spores) (Hammer et al. 1995) was detected as small pinpoint colonies on plate count agar (PCA) incubated at 30°C. This organism was described later as *B. sporothermodurans* (Pettersson et al. 1996). Even though it does not have any pathogenic or toxic activity (Hammer et al. 1995;Hammer and Walte 1996), and also only causes minor spoilage effects such as sometimes a slight pink discoloration (Klijn et al. 1997;Lembke 1995), contamination levels of 10^5 vegetative cells and 10^3 spores mL^{-1} milk far exceed the EC regulation. A molecular typing study of UHT-isolates from different countries as well as farm isolates suggested a clonal origin of the UHT-isolates (referred to as HRS-clone) (Guillaume-Gentil et al. 2002). This could probably be attributed to reprocessing and circulation of contaminated milk and the use of contaminated milk powder to reconstitute milk for UHT processing (Scheldeman et al. 2006). A less specific EC-regulation is now in place, stipulating microbiological stability of incubated UHT-cartons (Anonymous 2006).

Another issue that needs to be addressed when discussing aerobic spore-formers in the light of product quality, is bacteriological safety.

In 2008, ten EU-member states reported a total of 124 food-borne outbreaks caused by *Bacillus* spp. and two non-member states reported 9 *Bacillus* spp. outbreaks. Only 45 of the *Bacillus* outbreaks were verified (36.3%) with 1,132 cases; 41 cases were hospitalized. Compared to 2007 the total number of outbreaks caused by *Bacillus* spp. toxins within the EU had increased by 18.1%. *B. cereus* was identified as the causative agent in each of the verified cases (European Food Safety Authority 2010). This well known food pathogen can cause two types of food poisoning syndromes: (i) a diarrhoeal type, characterized by abdominal pain with diarrhoea 8 to 16 h after ingestion of the contaminated food and (ii) an emetic type that is characterized by nausea and vomiting and may even lead to fatalities (Dierick et al. 2005) with an onset 1 to 5 h after eating the affected food.

The diarrhoeal syndrome is associated with a diversity of foods such as meat, vegetable dishes, pastas, desserts, cakes, sauces and milk. It is caused by disruption of the integrity of the plasma membrane of epithelial cells by a variety of heat-labile protein enterotoxins, that are thought to be produced by vegetative cells in the small intestine itself (Granum 2002). The infective doses range from 10^4 – 10^9 cells per gram of food, depending on the proportion of spores present in the food to survive the acid barrier of the stomach (Logan 2004). Three pore-forming cytotoxins have been associated with diarrhoeal disease: two homologous three-component toxins and a single component cytotoxin (named haemolysin BL (Hbl), nonhaemolytic enterotoxin (Nhe) and cytotoxin K (CytK), respectively). At present, the relative importance of Hbl and Nhe in food poisoning is unknown, with Nhe being present in almost all tested *B. cereus*/*Bacillus thuringiensis* strains and Hbl in about 50% of them (Granum 2002). Two different forms of CytK have been described, the highly cytotoxic CytK-1 and the moderate CytK-2 variant, encoded by *cytK-1* and *cytK-2* genes, respectively (Fagerlund et al. 2004). The *cytK-1* gene has thus far only been detected in a limited number of *B. cereus* strains, that have been proposed to form a novel bacterial species, for which the name "*B. cytotoxis*" or "*B. cytotoxicus*" is suggested (Lapidus et al. 2008). The role of two other single-component proteinaceous enterotoxins in food poisoning, enterotoxin T (BceT) and enterotoxin FM (EntFM), has not yet been elucidated: BceT was absent in 57 out of 95 *B. cereus* strains and in 5 out of 7 strains involved in food poisoning and EntFM is a complete question mark, simply being cloned without any bacteriological characterization (Granum 2002).

The emetic syndrome is predominantly associated with the consumption of food rich in carbohydrates such as oriental rice dishes and pastas, though occasionally other foods (*e.g.,* pasteurized cream, milk pudding and reconstituted infant-feed formulas) can also be implicated (Logan 2004). It is caused by a small ring-shaped heat- and acid-stable dodecadepsipeptide named cereulide that is already produced in the food itself. About 10^5 – 10^8 cells per gram of food are required to form sufficient toxin (Logan 2004).

Psychrotolerant strains within the *B. cereus* group (belonging to the species *B. cereus s.s.* and *Bacillus weihenstephanensis* (Borge et al. 2001;Stenfors and Granum 2001)) usually are not associated with foodborne intoxications. However, psychrotolerant properties were detected in the causative agent (identified as *B. cereus*) of foodborne outbreaks in Spain and the Netherlands (Van Netten et al. 1990). Furthermore, a recent study by Thorsen et al. (2006) demonstrated cereulide production in two psychrotolerant *B. weihenstephanensis* strains at temperatures as low as 8°C.

Moreover, temperature abuse may already result in an outgrowth of mesophilic strains as demonstrated in a study by Odumeru et al. (1997) who detected enterotoxic activity upon moderate temperature abuse (10°C) of pasteurized milk that allowed growth of *B. cereus* in the range of 10^3 to 10^6 CFU mL^{-1}. Still, the practical relevance of these findings is yet to be validated.

Although *Bacillus* species other than *B. cereus* have been incriminated as food poisoning agents, the link between toxin production and foodborne illness has not been fully established. Increasing evidence for the production of both heat-stable and heat-labile toxins is becoming apparent through cellular assays that confirm both production and functionality of the toxins. An overview of species in which the presence of toxinogenic components has been detected, is shown in Table 5.

Most of these species are also found in milk, though at present no cases of food poisoning from consuming milk products has been reported. This table clearly shows that heat-sensitive and heat-stabile toxins outside the *B. cereus* group mostly belong to the *B. subtilis* group. The heat-stable toxins show a high resemblance with the physico-chemical characteristics of cereulide (high resista^2 nce to extreme heat, pH and enzymatic degradation) (From et al. 2005;Salkinoja-Salonen et al. 1999).

They have been characterised as surfactin isoforms, named lichenysin, pumilacidin, amylopsin and surfactin.

The surfactin superfamily is a family of structurally diverse, low molecular weight cyclic lactonic lipopeptides, that is well-known in strains of members of the *B. subtilis* group, where it represents one of the many types of antibiotics produced by this group of species.

Table 5. Overview of toxin-producing aerobic endospore-forming species outside the *Bacillus cereus* group. Numbers indicate the used assay: [1]cellular assays, [2]boar sperm cell motility inhibition assay, [3]PCR detection, [4]immunoassay kits. If determined, the identification of the toxinogenic components is represented by a letter: [a]lichenysin A, [b]pumilacidin, [c]amylosin, [d]surfactin. *isolated from food poisoning events, among others

Source	Heat resistant	Heat sensitive
Beattie and Williams (1999)[1,4]	*Br. brevis*	*Br. brevis*
	B. circulans	*B. circulans*
	B. subtilis	*B. subtilis*
	B. lentus	*B. lentus*
	B. licheniformis	*B. licheniformis*
Salkinoja-Salonen et al. (1999)[2,*]	*B. licheniformis*[a]	
Lindsey et al.(2000)[1]	*B. licheniformis*	
	B. pumilus	
Mikkola et al. (2000)[2,*]	*B. licheniformis*[a]	
Suominen et al. (2001)[2,*]	*B. pumilus*[b]	
Phelps and McKillip (2002)[3,4]	*B. amyloliquefaciens*	
Mikkola et al. (2004[2], 2007)	*B. amyloliquefaciens*[c]	
From et al. (2005)[1,2,3,4]	*B. licheniformis*	
	B. pumilus	
	B. subtilis	
Taylor et al. (2005)[1,*]	*B. licheniformis*	
	B. simplex	
	B. firmus	
	B. megaterium	
From et al. (2007a)[*]	*B. pumilus*[b]	
From et al. (2007b)[1,2]	*B. mojavensis*[d]	
Nieminen et al. (2007)[2]	*B. licheniformis*[a]	
	B. pumilus	
Apetroaie-Constantin et al. (2009)[1,2,*]	*B. subtilis*[c]	
	B. mojavensis[c]	
De Jonghe et al. (2010)[1]	*B. subtilis*	*B. subtilis*
	B. amyloliquefaciens	*B. pumilus*
		B. amyloliquefaciens

Fermentative Non Spore-Formers

Fermentative spoilage organisms – generally from the lactic acid producing group (*e.g.*, *Lactococcus*, *Lactobacillus*, *Leuconostoc*, *Enterococcus*, *Pediococcus* and *Streptococcus* (Chapman and Sharpe 1990)) or coliform group (mainly *Enterobacter* and *Klebsiella* (Suhren 1989)), are able to spoil raw milk and derived products when storage temperature throughout the cold chain was high enough to outgrow psychrotolerant bacteria or when the product composition is inhibitory to Gram-negative aerobic organisms.

LAB are normal inhabitants of the cow's teat and are also associated with silage and other animal feeds or feces. Coliform bacteria are present on the outside of the udder as a result of fecal contamination (Bramley and McKinnon 1990). Though LAB are mesophilic bacteria, undesired growth upon inadequate cooling can result in souring of fluid milk products due to production of small amounts of acetic and propionic acids (Shipe et al. 1978). A malty flavour results from the production of 3-methylbutanal by *Lactobacillus lactis* subsp. *lactis* biovar *maltigenes* (Morgan 1976). Production of extracellular polymers causes a ropy texture, usually traced back to specific strains of lactococci that produce a polysaccharide containing mainly glucose and galactose with small amounts of mannose, rhamnose and pentose (Cerning 1990;Cerning et al. 1992). Fermentation of lactose by LAB may also result in a sour taste and curdling of caseins when the milk is heated .

Various cheese defects can be attributed to gas formation by LAB or coliform bacteria: e.g., an "open" texture or fissures are linked to predominance of heterofermentative LAB (Lalaye et al. 1987), as well as gassy defects and white crystalline deposits in Cheddar cheese (Cromie et al. 1987;Rengpipat and Johnson 1989). Several defects in Mozarella cheese can be attributed to different *Lactobacillus* spp. (Hull et al. 1983;Hull et al. 1992). *L. delbrueckii* subsp. *bulgaricus* can cause a pink discoloration in some cheese varieties due to failure of lowering the redox-potential of the cheese (Shannon et al. 1969).

Early blowing is a gassy cheese defect that may occur when conditions of temperature and pH during manufacturing become favourable for growth of coliform bacteria. However, growth of coliform bacteria does not necessarily cause texture defects, because development of such defects depend on the ability of strains to ferment citric acid (e.g. *Enterobacter aerogenes*) (Walstra et al. 1999).

Flavour defects in Cheddar cheese can also result from growth of LAB: a fruity off-flavour can be attributed to production of esterase (usually *Lactococcus* spp.) (Bill et al. 1965), and phenolic flavour has been associated with *L. casei* subsp. *alactosus* and *L. casei* subsp. *rhamnosus* (Hull et al. 1992).

Loss of flavour in fermented milk products such as sour cream and cottage cheese, can occur when diacetyl is reduced to acetoin and 2,3-butanediol by lactococci, coliforms and yeasts (Frank and Marth 1988;Hogarty and Frank 1982;Wang and Frank 1981).

Yeasts and Molds

Since yeasts are able to grow well at low pH, they commonly cause fruity or yeasty odour and/or gas formation of fermented milk products such as cultured milks (e.g. yoghurt and butter milk) and fresh cheeses (e.g. cottage cheese), that provide a highly specialized ecological niche for yeasts that can use lactose or lactic acid and tolerate high salt concentrations (Fleet 1990). Yeasts that are able to produce proteolytic or lipolytic enzymes may also have a selective advantage in milk products, even those with low a_w.

Growth of spoilage molds on cheese is a problem that dates back to prehistory. Control measures such as pasteurization, added liquid smoke, the use of antimycotic chemicals and specialized packaging don't seem to be completely effective.

Scrutiny at the Spoilage Issue

When it comes to restraining bacteriological spoilage of milk and dairy products, it is important to know exactly what spoilage agent we are dealing with so that a justifiable course of action can be set up to limit the initial contamination and control the outgrowth of the responsible bacteria. This implicates not only a thorough identification of the responsible bacteria, but also a clear insight in the issue.

It is generally accepted that aerobe spore-formers are the main spoilers of pasteurized milk and dairy products, as their spores survive this heat treatment whereas *Pseudomonas*‘ thermoduric enzymes cause spoilage of milk and dairy products with a long shelf life (*i.e.,* UHT treated or powdered products) as the activity of *Pseudomonas* enzymes is thought to be too low to affect pasteurized milk during its shelf life. But is the spoilage issue really this straightforward? Figure 7 shows an overview of the complexity of the milk spoilage issue.

For pseudomonads, spoilage of heat treated milk and milk products can be attributed to heat-stable enzymes produced by vegetative cells in the raw milk, or to enzymes (both heat-stable and heat labile) that are secreted by *Pseudomonas* bacteria that entered the product due to post processing contamination. These two spoilage processes can take place in both UHT treated and pasteurized samples. However, it needs to be remarked that aseptic filling of UHT products limits the possibility of post processing contamination considerably. Although aerobe spore-formers are thought to be more important in spoilage of pasteurized products because of the supposed higher activity of their spoilage enzymes, *Pseudomonas* species that entered the milk through PPC will easily overgrow these organisms because of their much larger growth rates and much shorter lag phases under refrigeration temperatures (*i.e.*, the temperatures applied for storage of pasteurized products) (Sørhaug and Stepaniak 1997). Still, the importance of other genera might be underestimated as concluded by Nörnberg et al. (2010) who demonstrated marked proteolytic activity in strains of *Burkholderia, Klebsiella* and *Aeromonas*.

When it comes to aerobe spore-formers such as *Bacillus s.l.*, the generally accepted idea that they are the most important spoilers of pasteurized milk products because their spores survive pasteurization, originates from actual spoiled pasteurized dairy products from which *Bacillus* species, mainly *Bacillus cereus*, could be isolated (spoilage route n°1 in Figure 7).

However, studies on the heat resistance of their spoilage enzymes, indicated that they are equally resistant as *Pseudomonas* enzymes, able to withstand pasteurization and treatments applied during commercial milk powder manufacture (Chen et al. 2004) (spoilage route n°2 in Figure 7) (however, data on their resistance to UHT treatment are lacking (spoilage route n°3 in Figure 7)). This implicates that not only the spores that germinate upon these processing treatments can cause spoilage in the retail product, but that possibly also vegetative cells secrete thermoduric spoilage enzymes in the raw milk upon cold storage prior to heat treatment (Chen et al. 2004). The vegetative cells can be released from biofilms (in the milking equipment at the farm, in the pumping installation of the milk tanker and the pipelines in the dairy plant) or maybe a minor fraction from spores that were able to germinate upon cold storage of the raw milk - provided that the vegetative cells are psychrotolerant and therefore able to grow.

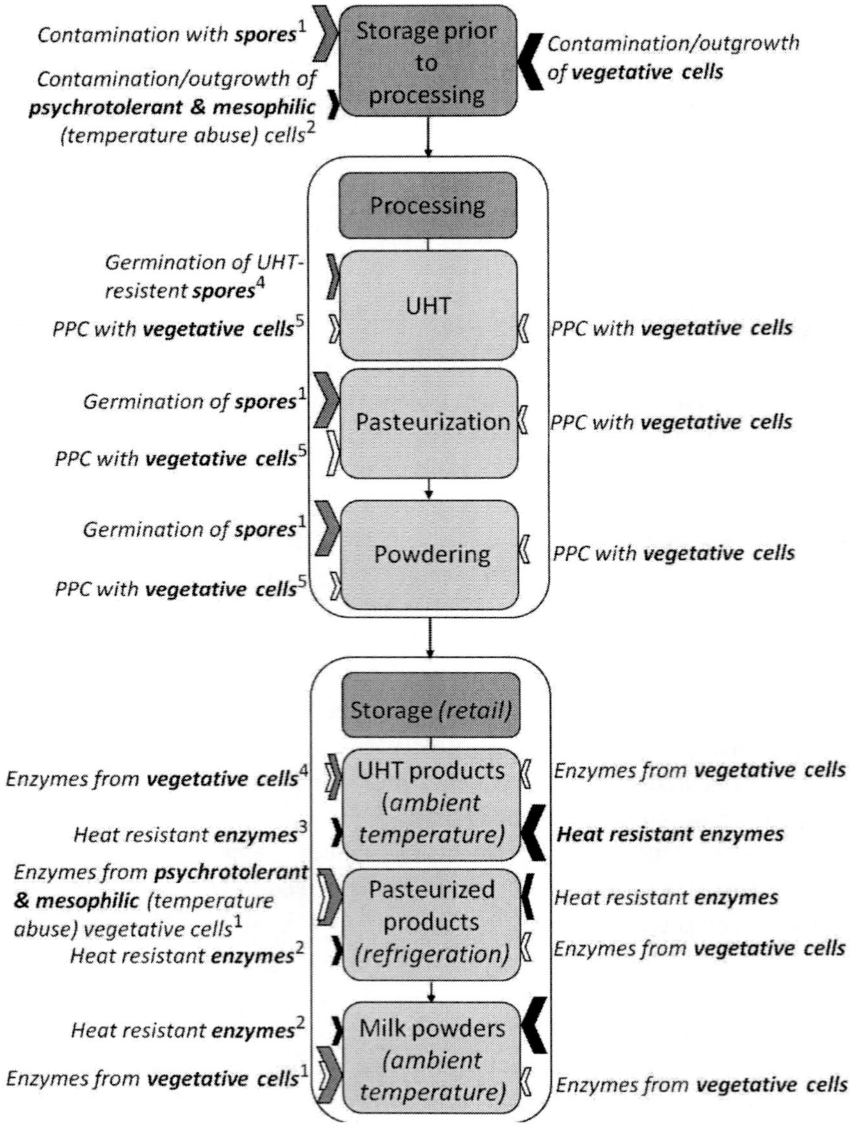

Figure 7. Flowchart of the milk spoilage issue. The interference points of *Bacillus s.l.* and *Pseudomonas* bacteria and spoilage enzymes are represented on the left and on the right, respectively. The relative importance of each step is reflected in the magnitude of the arrows. White arrows represent the spoilage issue caused by post processing contamination, whereas the black arrows represent spoilage issues caused by enzymes from vegetative cells that are already present prior to processing. For *Bacillus s.l.*, grey arrows represent spores already present prior to processing. PPC: post processing contamination.

As aerobe spore-formers tend to be present as spores in biofilms and generation times and lag phases of psychrotolerant *Bacillus s.l.* members at 2-7°C are considerably longer than those of *Pseudomonas* spp. (Chandler and McMeekin, 1985), this might nevertheless be a relatively less frequent phenomenon. However, under suboptimal storage temperatures, growth of vegetative *Bacillus* cells might be considerable. Indispensable knowledge to determine the importance of these spore-forming groups for spoilage of milk products may therefore be their possibility to grow out throughout the dairy chain (prior to processing). Although the predominant raw milk species *B. licheniformis*, *B. subtilis* and *B. pumilus* are generally regarded as mesophilic (Pacova et al. 2003), a fraction of their isolates show psychrotolerant traits that would enable them to grow out during cold storage of the raw milk. However, the presence of these strains in the form of vegetative cells has not yet been investigated in raw milk.

Furthermore, spores of certain *Bacillus s.l.* members can survive UHT treatment, thus ending up in a competition-free niche that is stored at room temperature, enabling them to grow out (and maybe produce spoilage enzymes) without restraints (spoilage route n°4 in Figure 7). However, except for *B. sporothermodurans*, no UHT-resistant bacilli have been linked to spoiled UHT-products up to now except in the event of PPC (a minor phenomenon because UHT milk is aseptically filled), which complicates the *Bacillus s.l.* spoilage route even more (spoilage route n°5 in Figure 7). Still, as UHT-products are required to be microbiologically stable upon storage at room temperature, outgrowth of HRS in itself might be enough to render these products unacceptable for consumption.

This raises the question as to the identity of the true culprit(s) when it comes to milk spoilage. Since growth rates of *Pseudomonas* members are much higher and lag phases much shorter at low storage temperatures of raw milk compared to these parameters in *Bacillus s.l.* species, it is likely that their production of heat resistant spoilage enzymes prior to processing will be much more significant. Nonetheless, the influence of *Bacillus'* heat resistant spoilage enzymes cannot simply be denied, certainly when storage temperatures rise to a suboptimal level (*e.g.,* 10°C), at which *Bacillus s.l.* species become the dominant microbiota (Sørhaug and Stepaniak 1997). This is also relevant for pasteurized milk products where spoilage enzymes from vegetative cells from both *Bacillus s.l.* and *Pseudomonas* members able to grow out in the end product after germination and PPC, respectively, are responsible for spoilage.

TIME FOR ACTION!

Zero Tolerance Policy: Reduction of the Bacterial Load of Raw Milk

An economically feasible solution for total elimination of milk spoilage is an illusion as this would require at least daily collection of the raw milk at the farm followed by immediate and intensive processing (using bactofugation, among others) at the dairy factory. But even though complete elimination of spores and psychrotolerant organisms in raw milk might not be feasible, this chapter would like to offer some perspective strategies to limit contamination with aerobe spore-formers and outgrowth of psychrotolerant bacteria.

Some aspects of farm management might take part in the contamination of raw milk with aerobe spore-formers. A study specifically for *B. cereus* estimated through predictive modelling that a 99% reduction in *B. cereus* spores could be achieved during the grazing period if soil contamination were minimized and teat cleaning were optimized. When the cows are housed, a 60% reduction of the *B. cereus* spore concentration should be feasible by ensuring spore concentrations in feed below 10^3 spores per gram and a pH of the ration offered to the cows below 5 (Vissers et al. 2007b). Also, Coorevits et al. (2008) revealed a somewhat different population structure within the total aerobe spore-forming microbiota in raw milk from different farm practices (organic versus conventional farming). They found a relatively higher number of thermotolerant organisms in milk from conventional dairy farms compared to organic farms (41.2% vs. 25.9%) and *B. cereus* group organisms and *Ureibacillus thermosphaericus* being predominant in organic and conventional milks, respectively. These differences could possibly be linked to differencest in housing and feeding strategies. Therefore, it was advised that future research should focus on specific contamination sources and concomitant advises for feed, pasturing and housing strategies.

Elimination strategies for aerobe spore-formers might entail induced germination of the spores just before processing (through the addition of germinant (mixtures) such as L-alanine and inosine (Hornstra et al. 2007)), after which the vegetative cells should be easily inactivated using commercially applied heat treatments. However, this apparently simple solution is hampered because germination of spore populations is very heterogeneous. Furthermore, some spores, known as superdormant, germinate extremely slowly (Ghosh and Setlow 2009).

As conventional heat treatment of milk such as pasteurization appears to be insufficient to kill off bacterial spores and certain spoilage enzymes, other (supplementary) techniques may be required to come to a microbiologically and enzymatically stable end product. High pressure (HP) homogenization (100-1000 megaPascals (MPa)) has the advantage that sensory and nutritional characteristics are generally unchanged (Thiebaud et al. 2003). Even though bacterial spores are highly pressure resistant, superdormant spores of *B. cereus* and *B. subtilis* appear to germinate just as well as dormant spores by pressures of 150 or 500 MPa (Wei et al. 2010). When nisin is added to the milk prior to HP treatment, the viability of spores may decrease even more (Black et al. 2008).

Also enzymes related to food quality can be deactivated by pressure, but the pressure needed strongly depends on the enzyme (Hendrickx et al. 1998). Therefore, a combined pressure-temperature treatment is the most appropriate approach for both pasteurization and sterilization processes (Hendrickx et al. 1998).

As for the psychrotolerant *Pseudomonas* microbiota from raw milk, the study by De Jonghe et al. (2011) demonstrated that the outgrowth and consequent production of (heat resistant) spoilage enzymes is not hampered by cooled storage of the raw milk as both suboptimally and optimally cooled milk supports growth of these psychrotolerant organisms. However, the effect of precooling of freshly obtained milk before it enters the farm tank was not yet taken into consideration. A few simple investments at the farm level such as precooling, preferentially with ice water, to eliminate milking peaks, adequate cooling throughout the entire cold chain and rapid processing of the raw milk at the dairy might make a world of difference in the prevention of outgrowth of pseudomonads in raw milk.

Alternatively, pressurized CO_2 (Werner and Hotchkiss 2006) or N_2 (Munsch-Alatossava et al. 2009) might represent relatively low-cost nonthermal methods that can be used in addition to commercially applied heat treatments to reduce microbial outgrowth of vegetative cells in raw milk.

A great number of spoilage microbiota are thought to originate from biofilms that are formed in cracks and cuts in the rubber materials of the milking equipment (Morse et al. 1968). The adherence to stainless steel, the recurrent flow of hot sanitizing chemicals and continuous flow of cold raw milk through these pipelines might create a selective platform for certain types of bacteria (Shaheen et al. 2010), which may explain the dominant *B. cereus* type that was found by De Jonghe et al. (2008). A possible strategy to limit the bacterial load in the raw milk might entail the use of silver-impregnated rubbers to avoid biofilm formation as silver is known for its anti-microbial properties (Sondi and Salopek-Sondi 2004).

If You Can't Beat Them, Detect Them to Avoid Them!

As a complete elimination of spoilage might be an utopia, a better approach to avoid economical losses is a proactive screen of the raw milk as it enters the dairy factory by a fast and easy detection method. Based on the detected spoilage potential, a better evaluation can be made of the shelf life of the end product or an appropriate processing method or destination can be chosen.

Molecular approaches are routinely used as screening methods because they are quick and easy to use. However, as more and more species are being identified within a certain genus, the taxonomic boundaries that lay in between them are becoming smaller. This implicates that much more sequence information is required to come to an unequivocal identification. Moreover, some old-school golden standards such as the 16S rRNA gene have become inadequate when it comes to identification at the species level in certain genera, especially *Pseudomonas*. This limits the use of nowadays popular molecular techniques such as 16S rDNA based DGGE and TGGE discrimination on a limited sequence variability. However, a combined approach of both cultivation and cultivation-independent methods resulting in the indication of representative marker strains might help to solve a lot of these issues (as proposed by De Jonghe et al. 2011). However, as farm management may play an important role in the composition of the raw milk (spoilage) microbiota (Coorevits et al. 2008), this approach might be management-specific.

Screening at the DNA-level always has the downside to it that the detected organism is not necessarily growing and actively producing spoilage enzymes. Furthermore, when screening for the presence of spoilage genes, the mere presence of the gene doesn't guarantee an active enzyme. These issues can be largely overcome by using a quantitative real time reverse transcriptase (RT)-PCR that detects the expression of the spoilage enzymes at the mRNA level. However, additional information is required that links the expression levels of spoilage enzymes to sensory and structural characteristics of the end product. Another more likely possibility is detection of spoilage enzymes at the protein level by means of ELISA (Enzyme Linked Immuno Sorbent Assay) which could be performed at the moment that a raw milk delivery is entering the dairy plant (*e.g.*, by means of a dipstick).

CONCLUSION

Though knowledge on the milk spoilage issue and the responsible bacteriological actors is growing, we are far from achieving an economically viable solution for it. Total eradication of the responsible bacteria seems an utopia, and then there's always the risk that the niche will be captured by other bacteria with unknown toxinogenic and spoilage capacities. In order to determine an appropriate destination and according processing conditions, detection of spoilage bacteria should take place very early-on in the raw milk, so that no irreversible damage to the end product has already been done.

REFERENCES

Abeni,F., Degano,L., Calza,F., Giangiacomo,R. and Pirlo,G. (2005) Milk quality and automatic milking: fat globule size, natural creaming, and lipolysis. *Journal of Dairy Science* 88, 3519-3529.

Adams,D.M., Barach,J.T. and Speck,M.L. (1975) Heat resistant proteases produced in milk by psychrotrophic bacteria of dairy origin. *Journal of Dairy Science* 58, 828-834.

Agata,N., Mori,M., Ohta,M., Suwan,S., Ohtani,I. and Isobe,M. (1994) A novel dodecadepsipeptide, cereulide, isolated from *Bacillus cereus* causes vacuole formation in HEp-2 cells. *FEMS Microbiology Letters* 121, 31-34.

Aires,G.S., Walter,E.H., Junqueira,V.C., Roig,S.M. and Faria,J.A. (2009) *Bacillus cereus* in refrigerated milk submitted to different heat treatments. *Journal of Food Protection* 72, 1301-1305.

Alichanidis,E., Wrathall,J.H.M. and Andrews,A.T. (1986) Heat stability of plasmin (milk proteinase) and plasminogen. *Journal of Dairy Research* 53, 259-269.

Andersson,A., Ronner,U. and Granum,P.E. (1995) What problems does the food industry have with the spore-forming pathogens *Bacillus cereus* and *Clostridium perfringens*? *International Journal of Food Microbiology* 28, 145-155.

Andersson,R.E., Hedlund,C.B. and Jonsson,U. (1979) Thermal inactivation of a heat resistant lipase produced by the psychrotrophic bacterium *Pseudomonas fluorescens*. *Journal of Dairy Science* 62, 361-367.

Anonymous (2006) Commission Regulation (EC) No 1662/2006 of 6 November 2006 amending Regulation (EC) No 853/2004 of the European Parliament and of the Council laying down specific rules for food of animal origin. *Official Journal of the European Union* L320, 1-10.

Apetroaie-Constantin,C., Mikkola,R., Andersson,M.A., Teplova,V., Suominen,I., Johansson,T. and Salkinoja-Salonen,M. (2009) *Bacillus subtilis* and *B. mojavensis* strains connected to food poisoning produce the heat stable toxin amylosin. *Journal of Applied Microbiology*.

Arpigny,J.L. and Jaeger,K.E. (1999) Bacterial lipolytic enzymes: classification and properties. *Biochemical Journal* 343, 177-183.

Banks, W., Dalgleish, D. G. and Rook, J. A. F. (1981) Milk and milk processing. In *Dairy Microbiology* ed. Robinson,R.K. London, UK: Applied Science.

Barkley,M.B. and Delaney,P. (1980) An occurence of bitty cream in Australian pasteurized milk. *Netherlands Milk and Dairy Journal* 34, 191-198.

Bastian,E.D. and Brown,R.J. (1996) Plasmin in milk and dairy products: an update. *International Dairy Journal* 6, 435-457.

Beattie,S.H. and Williams,A.G. (1999) Detection of toxigenic strains of *Bacillus cereus* and other *Bacillus* spp. with an improved cytotoxicity assay. *Letters in Applied Microbiology* 28, 221-225.

Bill,D.D., Morgan,M.E., Reddy,L.M. and Day,E.A. (1965) Identification of compounds responsible for fruit flavor defect of experimental Cheddar cheeses. *Journal of Dairy Science* 48, 1168-1170.

Billing,E. and Cuthbert,W.A. (1958) "Bitty cream": The occurence and significance of *Bacillus cereus* in raw milk supplies. *Journal of Applied Bacteriology* 21, 65-78.

Bishop,J.R. and White,C.H. (1986) Assessment of dairy product quality and potential shelf-life - a review. *Journal of Food Protection* 49, 739-748.

Black,E.P., Linton,M., McCall,R.D., Curran,W., Fitzgerald,G.F., Kelly,A.L. and Patterson,M.F. (2008) The combined effects of high pressure and nisin on germination and inactivation of *Bacillus* spores in milk. *Journal of Applied Microbiology* 105, 78-87.

Borge,G.I.A., Skeie,M., Sorhaug,T., Langsrud,T. and Granum,P.E. (2001) Growth and toxin profiles of *Bacillus cereus* isolated from different food sources. *International Journal of Food Microbiology* 69, 237-246.

Bramley, A. J. and McKinnon, C. H. (1990) The microbiology of raw milk. In *Dairy Microbiology. Volume 1* ed. Robinson,R.K. pp. 163-208. New York, USA: Elsevier Applied Science.

Braun,P., Fehlhaber,K., Klug,C. and Kopp,K. (1999) Investigations into the activity of enzymes produced by spoilage-causing bacteria: a possible basis for imprived shelf-life estimation. *Food Microbiology* 16, 531-540.

Bylund,G. (1995) The chemistry of milk. In *Dairy Processing Handbook* Lund, Sweden: Tetra Pak Processing Systems AB.

Cempirkova,R. (2002) Psychrotrophic vs. total bacterial counts in bulk milk samples. *Veterinarni Medicina* 47, 227-233.

Cerning,J. (1990) Exocellular polysaccharides produced by lactic acid bacteria. *FEMS Microbiology Review* 7, 113-130.

Cerning,J., Bouillanne,C., Landon,M. and Desmazeaud,M. (1992) Isolation and characterization of exopolysaccharides from slime-forming mesophilic lactic acid bacteria. *Journal of Dairy Science* 75, 692-699.

Chabeaud,P., de,G.A., Bitter,W., Tommassen,J., Heulin,T. and Achouak,W. (2001) Phase-variable expression of an operon encoding extracellular alkaline protease, a serine protease homolog, and lipase in *Pseudomonas brassicacearum*. *Journal of Bacteriology* 183, 2117-2120.

Champagne,C.P., Laing,R.R., Roy,D., Mafu,A.A. and Griffiths,M.W. (1994) Psychrotrophs in dairy products: their effects and their control. *Critical Reviews in Food Science and Nutrition* 34, 1-30.

Chandler,R.E. and McMeekin,T.A. (1985) Temperature function integration and its relationship to the spoilage of pasteurized, homogenized milk. *Australian Journal of Dairy Technology* 40, 37-41.

Chapman, H. R. and Sharpe, M. E. (1990) Microbiology of cheese. In *Dairy Microbiology. Volume 2* ed. Robinson,R.K. pp. 203-289. New York, USA: Elsevier Applied Science.

Chen,L., Coolbear,T. and Daniel,R.M. (2004) Characteristics of proteinases and lipases produced by seven *Bacillus* spp. isolated from milk powder production lines. *International Dairy Journal* 14, 495-504.

Chen,L., Daniel,R.M. and Coolbear,T. (2003) Detection and impact of protease and lipase activities in milk and milk powders. *International Dairy Journal* 13, 255-275.

Christiansson,A., Bertilsson,J. and Svensson,B. (1999) *Bacillus cereus* spores in raw milk: factors affecting the contamination of milk during the grazing period. *Journal of Dairy Science* 82, 305-314.

Cogan,T.M. (1977) A review of heat resistant lipases and proteinases and the quality of dairy products. *Irish Journal of Food Science and Technology* 1, 95-105.

Coorevits,A., De Jonghe,V., Vandroemme,J., Reekmans,R., Heyrman,J., Messens,W., De Vos,P. and Heyndrickx,M. (2008) Comparative analysis of the diversity of aerobic spore-forming bacteria in raw milk from organic and conventional dairy farms. *Systematic and Applied Microbiology* 31, 126-140.

Cousin,M.A. (1982) Presence and activity of psychrotrophic microorganisms in milk and dairy products: a review. *Journal of Food Protection* 45, 172-207.

Cousin, M. A. (1989) Physical and biochemical effects on milk components. In *Enzymes of Psychrotrophs in Raw Food* ed. McKellar,R.C. pp. 205-225. Boca Raton, Florida: CRC Press Inc.

Cousins, C. M. and Bramley, A. J. (1981) The microbiology of raw milk. In *Dairy Microbiology* ed. Robinson,R.K. pp. 119-163. London, UK: Applied Science.

Cromie,S.J., Giles,J.E. and Dulley,J.R. (1987) Effect of elevated ripening temperatures on the microflora of Cheddar cheese. *Journal of Dairy Research* 54, 69-76.

Crudden,A., Fox,P.F. and Kelly,A.L. (2005) Factors affecting the hydrolytic action of plasmin in milk. *International Dairy Journal* 15, 305-313.

Dasgupta,A.P. and Hull,R.R. (1989) Late blowing of swiss-cheese - incidence of *Clostridium tyrobutyricum* in manufacturing milk. *Australian Journal of Dairy Technology* 44, 82-87.

Datta,N. and Deeth,H.C. (2001) Age gelation of UHT-milk - a review. *Food And Bioproducts Processing* 79, 197-210.

Davies, F. L. and Wilkinson, G. (1973) *Bacillus* in milk and dairy products. In *The microbiological safety of food* ed. Hobbs,B.C. and Christian,J.H.B. pp. 57-67. London, United Kingdom: Acad. Press.

De Jonghe,V., Coorevits,A., De,B.J., Van,C.E., Grijspeerdt,K., Herman,L., De,V.P. and Heyndrickx,M. (2010) Toxinogenic and spoilage potential of aerobic spore-formers isolated from raw milk. *International Journal of Food Microbiology* 136, 318-325.

De Jonghe,V., Coorevits,A., De Vos,P. and Heyndrickx,M. (2011) Influence of storage conditions on the growth of *Pseudomonas* species in refrigerated raw milk. *Applied and Environmental Microbiology* 77, 460-470.

De Jonghe,V., Coorevits,A., Vandroemme,J., Heyrman,J., Herman,L., De Vos,P. and Heyndrickx,M. (2008) Intraspecific genotypic diversity of *Bacillus* species from raw milk. *International Dairy Journal* 18, 496-505.

Deeth,H.C. (2006) Lipoprotein lipase and lipolysis in milk. *International Dairy Journal* 16, 555-562.

Deeth, H. C. and Fitz-Gerald, C. H. (1983) Lipolytic enzymes and hydrolytic rancidity in milk and milk products. In *Developments in dairy chemistry. Part II* ed. Fox,P.F. pp. 195-239. London, UK: Applied Science.

Deeth, H. C. and Fitz-Gerald, C. H. (1994) Lipolytic enzymes and hydrolytic rancidity in milk and milk products. In *Advanced dairy chemistry, Vol. 2: Lipids* ed. Fox,P.F. pp. 247-308. London,UK: Chapman and Hall.

Derewenda,Z.S. and Derewenda,U. (1991) Relationships among serine hydrolases: evidence for a common structural motif in triacylglyceride lipases and esterases. *Biochemical and Cellular Biology* 69, 842-851.

Dierick,K., Van Coillie,E., Swiecicka,I., Meyfroidt,G., Devlieger,H., Meulemans,A., Hoedemaekers,G., Fourie,L., Heyndrickx,M. and Mahillon,J. (2005) Fatal family outbreak of *Bacillus cereus*-associated food poisoning. *Journal of Clinical Microbiology* 43, 4277-4279.

Dogan,B. and Boor,K.J. (2003) Genetic diversity and spoilage potentials among *Pseudomonas* spp. isolated from fluid milk products and dairy processing plants. *Applied and Environmental Microbiology* 69, 130-138.

Dufour,D., Nicodeme,M., Perrin,C., Driou,A., Brusseaux,E., Humbert,G., Gaillard,J.L. and Dary,A. (2008) Molecular typing of industrial strains of *Pseudomonas* spp. isolated from milk and genetical and biochemical characterization of an extracellular protease produced by one of them. *International Journal of Food Microbiology* 125, 188-196.

Eggert,T., Pencreac'h,G., Douchet,I., Verger,R. and Jaeger,K.E. (2000) A novel extracellular esterase from *Bacillus subtilis* and its conversion to a monoacylglycerol hydrolase. *European Journal of Biochemistry* 267, 6459-6469.

Eneroth,A., Christiansson,A., Brendehaug,J. and Molin,G. (1998) Critical contamination sites in the production line of pasteurised milk, with reference to the psychrotrophic spoilage flora. *International Dairy Journal* 8, 829-834.

European Food Safety Authority (2010) The community summary report on foodborne outbreaks in the European Union in 2008.

Fagerlund,A., Ween,O., Lund,T., Hardy,S.P. and Granum,P.E. (2004) Genetic and functional analysis of the *cytK* family of genes in *Bacillus cereus*. *Microbiology* 150, 2689-2697.

Fairbairn,D.J. and Law,B.A. (1986) Proteinases of psychrotrophic bacteria: their production, properties, effects and control. *Journal of Dairy Research* 53, 139-177.

Fajardo-Lira,C., Oria,M., Hayes,K.D. and Nielsen,S.S. (2000) Effect of psychrotrophic bacteria and of an isolated protease from *Pseudomonas fluorescens* M3/6 on the plasmin system of fresh milk. *Journal of Dairy Science* 83, 2190-2199.

Farkye, N. Y., Imafidon, G. I. and Fox, P. F. (1995) Thermal denaturation of indigenous milk enzymes. In *Heat induced changes in milk* ed. Fox,P.F. pp. 331-348. Brussels, Belgiuma: International Dairy Federation.

Fleet,G.H. (1990) Yeasts in dairy products. *Journal of Applied Bacteriology* 68, 199-211.

Fox,P.F. (1981) Proteinases in dairy technology. *Netherlands Milk and Dairy Journal* 35, 233-253.

Fox,P.F., Yaguchi,M. and Tarassuk,N.P. (1967) Distribution of lipase in milk proteins. II. Dissociation from κ-casein with dimethylformamide. *Journal of Dairy Science* 50, 307-312.

Frank, J. F. and Marth, E. H. (1988) Fermentations. In *Fundamentals of Dairy Chemistry* ed. Wong,N.P. pp. 656-738. New York, USA: Van Nostrand Reinhold Co.

From,C., Hormazabal,V. and Granum,P.E. (2007a) Food poisoning associated with pumilacidin-producing *Bacillus pumilus* in rice. *International Journal of Food Microbiology* 115, 319-324.

From,C., Hormazabal,V., Hardy,S.P. and Granum,P.E. (2007b) Cytotoxicity in *Bacillus mojavensis* is abolished following loss of surfactin synthesis: implications for assessment of toxicity and food poisoning potential. *International Journal of Food Microbiology* 117, 43-49.

From,C., Pukall,R., Schumann,P., Hormazabal,V. and Granum,P.E. (2005) Toxin-producing ability among *Bacillus* spp. outside the *Bacillus cereus* group. *Applied and Environmental Microbiology* 71, 1178-1183.

Garcia,M.L., Sanz,B., Garcia-Collia,P. and Ordonez,J.A. (1989) Activity and thermostability of the extracellular lipases and proteinases from pseudomonads isolated from raw milk. *Milchwissenschaft* 44, 547-560.

Gasincova,E., Stepaniak,L. and Sørhaug,T. (1994) Low temperature inactivation of lipase from *Pseudomonas*. *Italian Journal of Food Science* 6, 449-452.

Ghosh,S. and Setlow,P. (2009) Isolation and characterization of superdormant spores of *Bacillus* species. *Journal of Bacteriology* 191, 1787-1797.

Granum, P. E. (2002) *Bacillus cereus* and food poisoning. In *Applications and Systematics of Bacillus and Relatives* ed. Berkeley,R.C., Heyndrickx,M., Logan,N.A. and De Vos,P. pp. 37-46. Oxford, UK.: Blackwell Science.

Griffiths,M.W. (1983) Synergistic effects of various lipases and phospholipase C on milk fat. *Journal of Food Technology* 18, 495-505.

Griffiths,M.W., Phillips,J.D. and Muir,D.D. (1981) Thermostability of proteases and lipases from a number of species of psychrotrophic bacteria of dairy origin. *Journal of Applied Bacteriology* 50, 289-303.

Grufferty,M.B. and Fox,P.F. (1988) Milk alkaline proteinase. *Journal of Dairy Research* 55, 609-630.

Guillaume-Gentil,O., Scheldeman,P., Marugg,J., Herman,L., Joosten,H. and Heyndrickx,M. (2002) Genetic heterogeneity in *Bacillus sporothermodurans* as demonstrated by ribotyping and repetitive extragenic palindromic-PCR fingerprinting. *Applied and Environmental Microbiology* 68, 4216-4224.

Guinot-Thomas,P., Al Ammoury,M., Le Roux,Y. and Laurent,F. (1995) Study of proteolysios during storage of raw milk at 4°C: effect of plasmin and microbial proteinases. *International Dairy Journal* 5, 685-697.

Gupta,R., Gupta,N. and Rathi,P. (2004) Bacterial lipases: an overview of production, purification and biochemical properties. *Applied Microbiology and Biotechnology* 64, 763-781.

Hammer,P., Lembke,F., Suhren,G. and Heeschen,W. (1995) Characterization of a heat resistant mesophilic *Bacillus* species affecting quality of UHT-milk - a preliminary report. *Kieler Milchwirtschaftliche Forschungberichte* 47, 303-311.

Hammer,P. and Walte,H.G. (1996) Zur pathogenität hitzeresistenter mesophiler sporenbilder aus UHT-milch. *Kieler Milchwirtschaftliche Forschungberichte* 48, 151-161.

Hanson,M.L., Wendorff,W.L. and Houck,K.B. (2005) Effect of heat treatment of milk on activation of *Bacillus* spores. *Journal of Food Protection* 68, 1484-1486.

Hardham,J.F. (1998) Effect of protein standardization of milk by addition of UF milk permeate on the composition and storage stability of UHT processed milk. *Australian Journal of Dairy Technology* 53, 22-27.

Harwalker, V. R. (1982) Age gelation of sterilized milks. In *Development in Dairy Chemistry. Volume 1* ed. Fox,P.F. pp. 229-269. London, UK: Elsevier Science Publisher Ltd.

Hayes, M. C. and Boor, K. J. (2001) Raw milk and fluid products. In *Applied Dairy Microbiology* ed. Marth,E.H. and Steele,J.L. pp. 59-76. New York: Marcel Dekker.

Hendrickx,M., Ludikhuyze,L., Van den broeck,I. and Weemaes,C. (1998) Effects of high pressure on enzymes related to food quality. *Trends in Food Science and Technology* 9, 197-203.

Heyndrickx, M. and Scheldeman, P. (2002) Bacilli associated with spoilage in dairy and other food products. In *Applications and Systematics of Bacillus and Relatives* ed. Berkely,R., Heyndrickx,M., Logan,N.A. and De Vos,P. pp. 64-82. Oxford, UK: Blackwell Science.

Hogarty,S.L. and Frank,J.F. (1982) Low-temperature activity of lactic streptococci isolated from cultured buttermilk. *Journal of Food Protection* 43, 1208-1211.

Hornstra,L.M., de Leeuw,P.L., Moezelaar,R., Wolbert,E.J., de Vries,Y.P., de Vos,W.M. and Abee,T. (2007) Germination of *Bacillus cereus* spores adhered to stainless steel. *International Journal of Food Microbiology* 116, 367-371.

Huemer,I.A., Klijn,N., Vogelsang,H.W.J. and Langeveld,L.P.M. (1998) Thermal death kinetics of spores of *Bacillus sporothermodurans* isolated from UHT milk. *International Dairy Journal* 8, 851-855.

Hull,R.R., Roberts,A.V. and Mayes,J.J. (1983) The association of *Lactobacillus casei* with a soft-body defect in commercial Mozarella cheese. *Australian Journal of Dairy Technology* 22, 78-80.

Hull,R.R., Toyne,S., Haynes,I. and Lehman,F. (1992) Thermoduric bacteria: a re-emerging problem in cheese making. *Australian Journal of Dairy Technology* 47, 91-94.

Hurley,M.J., Larsen,L.B., Kelly,A.L. and McSweeney,P.L.H. (2000) The milk acid proteinase cathepsin D: a review. *International Dairy Journal* 10, 673-681.

IDF (1987) Significance of lipolysis in the manufacture and storage of dairy products. In *Bulletin No. 144* Brussels, Belgium: International Dairy Federation.

Jaeger,K.E. and Eggert,T. (2002) Lipases for biotechnology. *Current Opinion in Microbiology* 13, 390-397.

Jaeger,K.E., Ransac,S., Dijkstra,B.W., Colson,C., van Heuvel,M. and Misset,O. (1994) Bacterial lipases. *FEMS Microbiology Review* 15, 29-63.

Jayarao,B.M. and Wang,L. (1999) A study on the prevalence of gram-negative bacteria in bulk tank milk. *Journal of Dairy Science* 82, 2620-2624.

Kalogridou-Vassiliadou,D. (1992) Biochemical activities of *Bacillus* species isolated from flat sour evaporated milk. *Journal of Dairy Science* 75, 2681-2686.

Kelly,A.L., O'Flaherty,F. and Fox,P.F. (2006) Indigenous proteolytic enzymes in milk: a brief overview of the present state of knowledge. *International Dairy Journal* 16, 563-572.

Klei,L.R., Lynch,J.M., Barbano,D.M., Oltenacu,P.A., Lednor,A.J. and Bandler,D.K. (1997) Influence of milking three times a day on milk quality. *Journal of Dairy Science* 80, 427-436.

Klijn,N., Herman,L., Langeveld,M.A., Vaerewijck,M. and Waegendorp,A.A. (1997) Genotypical and phenotypical characterization of *Bacillus* strains, surviving UHT sterilisation. *International Dairy Journal* 7, 421-428.

Klijn,N., Nieuwenhof,F.F., Hoolwerf,J.D., van der Waals,C.B. and Weerkamp,A.H. (1995) Identification of *Clostridium tyrobutyricum* as the causative agent of late blowing in cheese by species-specific PCR amplification. *Applied and Environmental Microbiology* 61, 2919-2924.

Kohlmann,K.L., Nielsen,S.S., Steenson,L.R. and Ladisch,M.R. (1991) Production of proteases by psychrotrophic microorganisms. *Journal of Dairy Science* 74, 3275-3283.

Koka,R.A. and Weimer,B.C. (2001) Influence of growth conditions on heat-stable phospholipase activity in *Pseudomonas*. *Journal of Dairy Research* 68, 109-116.

Kolar,C.W.Jr. and Mickle,J.B. (1963) Relationships between milk fat acidity, short-chain fatty acids and rancid flavours in milk. *Journal of Dairy Science* 46, 569-571.

Kroll, S. (1989) Thermal stability. In *Enzymes of psychrotrophs in raw food* ed. McKellar,R.C. pp. 121-152. Boca Raton, Florida: CRC Press, Inc.

Kumeta,H., Hoshino,T., Goda,T., Okayama,T., Shimada,T., Ohgiya,S., Matsuyama,H. and Ishizaki,K. (1999) Identification of a member of the serralysin family isolated from a psychrotrophic bacterium, *Pseudomonas fluorescens* 114. *Bioscience, Biotechnology and Biochemistry* 63, 1165-1170.

Kurzweil,R. and Busse,M. (1973) Keimgehalt und florazusammenzetsung der frisch ermolkenen milch. *Milchwissenschaft* 28, 427-431.

Labots,H. and Hup,G. (1964) *Bacillus cereus* in raw and pasteurized milk. I. Enumeration of *B. cereus* spores in milk. *Netherlands Milk and Dairy Journal* 19, 112-126.

Lafarge,V., Ogier,J.C., Girard,V., Maladen,V., Leveau,J.Y., Gruss,A. and Delacroix-Buchet,A. (2004) Raw cow milk bacterial population shifts attributable to refrigeration. *Applied and Environmental Microbiology* 70, 5644-5650.

Lalaye,L.C., Simard,R.E., Lee,B.H., Holley,R.A. and Giroux,R.N. (1987) Involvement of heterofermentative lactobacilli in development of open texture in cheeses. *Journal of Food Protection* 50, 1009-1012.

Lapidus,A., Goltsman,E., Auger,S., Galleron,N., Segurens,B., Dossat,C., Land,M.L., Broussolle,V., Brillard,J., Guinebretiere,M.H., Sanchis,V., Nguen-The,C., Lereclus,D., Richardson,P., Wincker,P., Weissenbach,J., Ehrlich,S.D. and Sorokin,A. (2008) Extending the *Bacillus cereus* group genomics to putative food-borne pathogens of different toxicity. *Chemico-Biological Interactions* 171, 236-249.

Law,B.A. (1979) Reviews in the progress of dairy science: enzymes of psychrotrophic bacteria and their effects on milk and milk products. *Journal of Dairy Research* 46, 573-588.

Law,B.A., Andrews,A.T. and Sharpe,M.E. (1977) Gelation of UHT-sterilized milk by proteases from a strain of *Pseudomonas fluorescens*, isolated from raw milk. *Journal of Dairy Research* 44, 145-148.

Le Bourhis,A.G., Saunier,K., Dore,J., Carlier,J.P., Chamba,J.F., Popoff,M.R. and Tholozan,J.L. (2005) Development and validation of PCR primers to assess the diversity of Clostridium spp. in cheese by temporal temperature gradient gel electrophoresis. *Applied and Environmental Microbiology* 71, 29-38.

Lembke, F. (1995) Highly heat resistant spores in UHT-milk. In *Bulletin No.302* pp. 60-61. Brussels, belgium: International Dairy Federation.

Liao,C.H. and McCallus,D.E. (1998) Biochemical and genetic characterization of an extracellular protease from *Pseudomonas fluorescens* CY091. *Applied and Environmental Microbiology* 64, 914-921.

Lindsay,D., Mosupye,F.M., Brozel,V.S. and von Holy,A. (2000) Cytotoxicity of alkaline-tolerant dairy-associated *Bacillus* spp. *Letters in Applied Microbiology* 30, 364-369.

Logan, N. A. (2004) Safety of aerobic endospore-forming bacteria. In *Bacterial spore formers. probiotics and emerging applications* ed. Ricca,E., Henriques,A.O. and Cutting,S.M. pp. 93-105. Norfolk, UK: Horizon Bioscience.

Lukasova,J., Vyhnalkova,J. and Pacova,Z. (2001) *Bacillus* species in raw milk and in the farm environment. *Milchwissenschaft* 56, 609-611.

Mabbit,L.A. (1981) The significant metabolic activity of bacteria in raw milk. *Kieler Milchwirtschaftliche Forschungberichte* 33, 273-280.

Macrae, A. R. (1983) Extracellular microbial lipases. In *Microbial enzymes and biotechnology* ed. Fogarty,W.M. pp. 225-249. New York, USA: Applied Science Publishers.

Magnusson,M., Christiansson,A. and Svensson,B. (2007) *Bacillus cereus* spores during housing of dairy cows: factors affecting contamination of raw milk. *Journal of Dairy Science* 90, 2745-2754.

Mansbridge,R.J. and Blake,J.S. (1997) Nutritional factors affecting the fatty acid composition of bovine milk. *British Journal of Nutrition* 78, S37-S47.

Månsson,H.L. (2008) Fatty acids in bovine milk fat. *Food and Nutrition Research* 52, DOI -10.3402/fnr.v52i0.1821.

Marchand,S., Coudijzer,K., Heyndrickx,M., Dewettinck,K. and De Block,J. (2008) Selective determination of the heat resistant proteolytic activity of bacterial origin in raw milk. *International Dairy Journal* 18, 514-519.

Marchand,S., Heylen,K., Messens,W., Coudijzer,K., De Vos,P., Dewettinck,K., Herman,L., De Block,J. and Heyndrickx,M. (2009a) Seasonal influence on heat resistant proteolytic capacity of *Pseudomonas lundensis* and *Pseudomonas fragi*, predominant milk spoilers isolated from Belgian raw milk samples. *Environmental Microbiology* 11, 467-482.

Marchand,S., Vandriesche,G., Coorevits,A., Coudijzer,K., De Jonghe,V., Dewettinck,K., De Vos,P., Devreese,B., Heyndrickx,M. and De Block,J. (2009b) Heterogeneity of heat resistant proteases from milk *Pseudomonas* species. *International Journal of Food Microbiology* 133, 68-77.

Marin,A., Mawhinney,T.P. and Marshall,R.T. (1984) Glycosidic activities of *Pseudomonas fluorescens* on fat-extracted skim milk, buttermilk, and milk fat globule membranes. *Journal of Dairy Science* 67, 52-59.

McCarthy,C.N., Woods,R.G. and Beacham,I.R. (2004) Regulation of the *aprX-lipA* operon of *Pseudomonas fluorescens* B52: differential regulation of the proximal and distal genes, encoding protease and lipase, by *ompR-envZ*. *FEMS Microbiology Letters* 241, 243-248.

McKellar,R.C. (1982) Factors influencing the production of extracellular proteinase by *Pseudomonas fluorescens*. *Journal of Applied Bacteriology* 53, 305-316.

McKellar, R. C. (1989) Regulation and control of synthesis. In *Enzymes of Psychrotrophs in Raw Food* ed. McKellar,R.C. pp. 153-172. Boca Raton, Florida: CRC Press, Inc.

McKellar,R.C., Froehlich,D.A., Butler,G., Cholette,H. and Campbell,C. (1984) The effect of uncooled storage on proteolysis, bitterness and apparent viscosity in UHT milk. *Canadian Institute of Food Science and Technology Journal* 17, 14-17.

McMahon,D.J. and Brown,R.J. (1984) Composition, structure and integrity of casein micelles: a review. *Journal of Dairy Science* 67, 499-512.

Meer,R.R., Bakker,J., Bodyfelt,F.W. and Griffiths,M.W. (1991) Psychotrophic *Bacillus* spp. in fluid milk products: a review. *Journal of Food Protection* 54, 969-979.

Metwalli,A.A.M., de Hongh,H.H.J. and van Boekel,M.A.J.S. (1998) Heat inactivation of bovine plasmin. *International Dairy Journal* 8, 47-56.

Mikkola,R., Andersson,M.A., Grigoriev,P., Teplova,V.V., Saris,N.E., Rainey,F.A. and Salkinoja-Salonen,M.S. (2004) *Bacillus amyloliquefaciens* strains isolated from moisture-damaged buildings produced surfactin and a substance toxic to mammalian cells. *Archives of Microbiology* 181, 314-323.

Mikkola,R., Andersson,M.A., Teplova,V., Grigoriev,P., Kuehn,T., Loss,S., Tsitko,I., Apetroaie,C., Saris,N.E., Veijalainen,P. and Salkinoja-Salonen,M.S. (2007) Amylosin from *Bacillus amyloliquefaciens*, a K+ and Na+ channel-forming toxic peptide containing a polyene structure. *Toxicon* 49, 1158-1171.

Mikkola,R., Kolari,M., Andersson,M.A., Helin,J. and Salkinoja-Salonen,M.S. (2000) Toxic lactonic lipopeptide from food poisoning isolates of *Bacillus licheniformis*. *European Journal of Biochemistry* 267, 4068-4074.

Morgan,M.E. (1976) The chemistry of some microbially induced flavor defects in milk and dairy foods. *Biotechnology and Bioengineering* 18, 953-965.

Morihara,K. (1974) Comparative specificity of microbial proteinases. *Advances in Enzymology* 41, 179-242.

Morse,P.M., Jackson,H., McNaughton,C.H., Leggat,A.G., Landerkin,G.B. and Johns,C.K. (1968) Investigation of factors contributing to the bacteria count of bulk tank milk. II. bacteria in milk from individual cows. *Journal of Dairy Science* 51, 1188-1191.

Mottar, J. (1989) Effect on the quality of dairy products. ed. McKellar,R.C. pp. 227-243. Boca Raton, Florida: CRC Press Inc.

Mottar,J., Waes,G., Moermans,R. and Naudts,M. (1979) Sensoric changes in UHT milk during uncooled storage. *Milchwissenschaft* 34, 257-262.

Mozaffarian,D., Aro,A. and Willett,W.C. (2009) Health effects of trans-fatty acids: experimental and observational evidence. *Eur. J. Clin. Nutr.* 63 Suppl 2, S5-21.

Muir, D. D. (1989) The microbiology of heat treated fluid milk products. In *Dairy Microbiology. Volume 1* ed. Robinson,R.K. pp. 209-270. New York, USA: Elsevier Applied Science.

Munsch-Alatossava,P., Gursoy,O. and Alatossava,T. (2009) Potential of nitrogen gas (N_2) to control psychrotrophs and mesophiles in raw milk. *Microbiological Research* 165, 122-132.

Nahsif,S.A. and Nelson,F.E. (1953) The lipase of *Pseudomonas fragi*. III. Enzyme action in cream and butter. *Journal of Dairy Science* 36, 481-488.

Ney, K. H. (1979) Bitterness of peptides: amino acid composition and chain length. In *Food Taste Chemistry, ACS symposium series 115* ed. Boudreau,J.C. pp. 149-173. Washington: American Chemical Society.

Nieminen,T., Rintaluoma,N., Andersson,M., Taimisto,A.M., Ali-Vehmas,T., Seppala,A., Priha,O. and Salkinoja-Salonen,M. (2007) Toxinogenic *Bacillus pumilus* and *Bacillus licheniformis* from mastitic milk. *Veterinary Microbiology* 124, 329-339.

Nörnberg,M., Friedrich,R., Weis,R., Tondo,E. and Brandelli,A. (2010) Proteolytic activity among psychrotrophic bacteria isolated from refrigerated raw milk. *International Journal of Dairy Technology* 63, 41-46.

Novak,J.S., Call,J., Tomasula,P. and Luchansky,J.B. (2005) An assessment of pasteurization treatment of water, media, and milk with respect to *Bacillus* spores. *Journal of Food Protection* 68, 751-757.

Odumeru,J.A., Toner,A.K., Muckle,C.A., Griffiths,M.W. and Lynch,J.A. (1997) Detection of *Bacillus cereus* diarrheal enterotoxin in raw and pasteurized milk. *Journal of Food Protection* 60, 1931-1393.

Olivecrona, T. (1980) Biochemical aspects of lipolysis in bovine milk. In *Bulletin No. 118* pp. 19-25. Brussels, Belgium: International Dairy Federation.

Olivecrona, T., Vilaro, S. and Olivecrona, G. (2003) Lipases in milk. In *Advanced dairy chemistry, Vol. 1. Proteins* ed. Fox,P.F. and McSweeney,P.L.H. pp. 473-494. New York, USA: Kluwer Academic Publishers/Plenum Press.

Overcast,W.W. and Atmaram,K. (1974) The role of *Bacillus cereus* in sweet curdling of fluid milk. *Journal of Milk and Food Technology* 37, 233-236.

Owens,J.J. (1978) Lecithinase positive bacteria in milk. *Process Biochemistry* 13, 13-15.

Pacova,Z., Svec,P., Stenfors,L.P., Vyletelova,M. and Sedlacek,I. (2003) Isolation of the psychrotolerant species *Bacillus weihenstephanensis* from raw cow's milk. *Czech Journal of Animal Science* 48, 93-96.

Palmer,J. (1981) Contamination of milk from the milking evironment. *Kieler Milchwirtschaftliche Forschungberichte* 33, 307-316.

Pettersson,B., Lembke,F., Hammer,P., Stackebrandt,E. and Priest,F.G. (1996) *Bacillus sporothermodurans*, a new species producing highly heat resistant endospores. *International Journal of Systematic Bacteriology* 46, 759-764.

Phadungath,C. (2005) Casein micelle structure: a conscise review. *Journal of Science and Technology* 27, 201-212.

Phelps,R.J. and McKillip,J.L. (2002) Enterotoxin production in natural isolates of *Bacillaceae* outside the *Bacillus cereus* group. *Applied and Environmental Microbiology* 68, 3147-3151.

Phillips,J.D. and Griffiths,M.W. (1986) Factors contributing to the seasonal variation of *Bacillus* spp. in pasteurized dairy products. *Journal of Applied Bacteriology* 61, 275-285.

Prado,B.M., Sombers,S.E., Ismail,B. and Hayes,K.D. (2006) Effect of heat treatment on the activity of inhibitors of plasmin and plasminogen activators in milk. *International Dairy Journal* 16, 593-599.

Priest,F.G. (1977) Extracellular enzyme synthesis in the genus *Bacillus*. *Bacteriological Reviews* 41, 711-753.

Quiberoni,A., Guglielmotti,D. and Reinheimer,J. (2008) New and classical spoilage bacteria causing widespread blowing in Argentinean soft and semihard cheeses. *International Journal of Dairy Technology* 61, 358-363.

Ranieri,M.L., Huck,J.R., Sonnen,M., Barbano,D.M. and Boor,K.J. (2009) High temperature, short time pasteurization temperatures inversely affect bacterial numbers during refrigerated storage of pasteurized fluid milk. *Journal of Dairy Science* 92, 4823-4832.

Rawlings,N.D. and Barrett,A.J. (1995) Evolutionary families of metallopeptidases. *Methods in Enzymology* 248, 183-228.

Rengpipat,S. and Johnson,E.A. (1989) Characterization of a *Lactobacillus* strain producing white crystals on cheddar cheese. *Applied and Environmental Microbiology* 55, 2579-2582.

Richard,J. (1981) Influence de diverses methodes de nettoyage des machines à traire sur la "qualité de conservation" du lail cru à basse température. *Lait* 61, 354-369.

Rollema,H.S., Visser,S. and Poll,J.K. (1981) On the determination, purification and characterization of the alkaline proteinase from bovine milk. *Netherlands Milk and Dairy Journal* 35, 396-399.

Rombaut,R., Dewettinck,K., de Mangelaere,G., van Vooren,L. and Huyghebaert,A. (2002) Raw milk microbial quality and production scale of Belgian dairy farms. *Milchwissenschaft* 57, 625-628.

Rysstad,G. and Kolstad,J. (2006) Extended shelf life milk - advances in technology. *International Journal of Dairy Technology* 59, 85-96.

Salkinoja-Salonen,M.S., Vuorio,R., Andersson,M.A., Kampfer,P., Andersson,M.C., Honkanen-Buzalski,T. and Scoging,A.C. (1999) Toxigenic strains of *Bacillus licheniformis* related to food poisoning. *Applied and Environmental Microbiology* 65, 4637-4645.

Scanlan,R.A., Sather,L.A. and Day,E.A. (1965) Contribution of free fatty acids to the flavor of rancid milk. *Journal of Dairy Science* 48, 1582-1584.

Scheldeman,P., Goossens,K., Rodriguez-Diaz,M., Pil,A., Goris,J., Herman,L., De Vos,P., Logan,N.A. and Heyndrickx,M. (2004) *Paenibacillus lactis* sp. nov., isolated from raw and heat-treated milk. *International Journal of Systematic and Evolutionary Microbiology* 54, 885-891.

Scheldeman,P., Herman,L., Foster,S. and Heyndrickx,M. (2006) *Bacillus sporothermodurans* and other highly heat resistant spore formers in milk. *Journal of Applied Microbiology* 101, 542-555.

Scheldeman,P., Pil,A., Herman,L., De Vos,P. and Heyndrickx,M. (2005) Incidence and diversity of potentially highly heat resistant spores isolated at dairy farms. *Applied and Environmental Microbiology* 71, 1480-1494.

Schröder,M.J.A. (1984) Origin and levels of post pasteurization contamination of milk in the dairy industry and the effects on the keeping quality. *Journal of Dairy Research* 51, 59-67.

Shaheen,R., Svensson,B., Andersson,M.A., Christiansson,A. and Salkinoja-Salonen,M. (2010) Persistence strategies of *Bacillus cereus* spores isolated from dairy silo tanks. *Food Microbiology* 27, 347-355.

Shamsuzzaman,K., Moddler,W. and McKellar,R.C. (1989) Survival of lipase during manufacture of nonfat dry milk. *Journal of Dairy Science* 70, 746-751.

Shannon,E.L., Olson,N.F. and von Elbe,J.H. (1969) Effect of lactic starter culture on pink discoloration and oxidation-reduction potential in Italian cheese. *Journal of Dairy Science* 52, 1567-1571.

Shehata,T.E., Duran,A. and Collins,E.B. (1971) Influence of temperature on the growth of psychrophilic strains of *Bacillus*. *Journal of Dairy Science* 54, 1579-1582.

Shelley,A.W., Deeth,H.C. and MacRae,I.C. (1986) Growth of lipolytic psychrotrophic pseudomonads in raw and ultra-heat-treated milk. *Journal of Applied Bacteriology* 61, 395-400.

Shelley,A.W., Deeth,H.C. and MacRae,I.C. (1987) A numerical taxonomic study of psychrotrophic bacteria associated with lipolytic spoilage of raw milk. *Journal of Applied Bacteriology* 62, 197-207.

Shinde,U. and Inouye,M. (1993) Intramolecular chaperones and protein folding. *Trends in Biochemical Science* 18, 442-446.

Shipe,W.F., Basette,R., Deane,D.D., Dunkley,W.L., Hammond,E.G., Harper,W.J., Kleyn,D.H., Morgan,M.E., Nelson,J.H. and Scanlan,R.A. (1978) Off flavors in milk: nomenclature, standards and bibliography. *Journal of Dairy Science* 61, 856-869.

Silanikove, N. (2008) Bioactive components of milk. In *Advances in Experimental Medicine and Biology. Volume 606* ed. Bösze,S. pp. 143-161. New York, USA: Springer.

Silanikove,N., Merin,U. and Leitner,G. (2006) Physiological role of indigenous milk enzymes: an overwiew of an evolving picture. *International Dairy Journal* 16, 533-545.

Simões,M., Simões,L.C. and Vieira,M.J. (2009) Species association increases biofilm resistance to chemical and mechanical treatments. *Water Research* 43, 229-237.

Slaghuis, B. A., Bos, K., De Jong, O., Tudos, A. J., te Giffel, M. C. and de Koning, K. (2004) Robotic milking and free fatty acids. In *Automatic milking. A better understanding* ed. Meijering,A., Hogeveen,H. and de Koning,C.J.A.M. pp. 341-347. Wageningen, The Netherlands: Academic Publishing.

Søgaard, H. and Lund, R. (1981) Psychrotrophic counts in raw milk and the keeping quality of pasteurized milk products. In *Psychrotrophic microorganisms in spoilage and pathogenicity* ed. Roberts,T.A., Hobbs,G., Christian,N. and Skøvgaard,N. pp. chapter 8. London, UK: Academic Press.

Sondi,I. and Salopek-Sondi,B. (2004) Silver nanoparticles as antimicrobial agent: a case study on *E. coli* as a model for Gram-negative bacteria. *Journal of Colloid and Interface Science* 275, 177-182.

Sørhaug,T. and Stepaniak,L. (1997) Psychrotrophs and their enzymes in milk and dairy products: quality aspects. *Trends in Food Science and Technology* 8, 35-41.

Speck, R. V. (1976) Aciduric flat sour sporeformes. In *Compendium of Methods for the Microbiological Examination of Foods* ed. Speck,M.L. pp. 248. Washington, DC: American Public Health Association, Inc.

Stadhouders, J. and Jørgensen, K. (1990) Prevention of the contamination of raw milk by a hygienic milk production. In *Bulletin No. 251* pp. 32-36. Brussels, Belgium: International Dairy Federation.

Stead,D. (1986) Microbial lipases: their characteristics, role in food spoilage and indsutrial uses. *Journal of Dairy Research* 53, 481-505.

Stenfors,L.P. and Granum,P.E. (2001) Psychrotolerant species from the *Bacillus cereus* group are not necessarily *Bacillus weihenstephanensis*. *FEMS Microbiology Letters* 197, 223-228.

Stepaniak,L. (1991) Factors affecting quality and possibilities of predicting shelf-life of pasteurized and ultra-high temperature heated milks. *Italian Journal of Food Science* 4, 11-26.

Stewart,D.B. (1975) Factors influencing the incidence of *Bacillus cereus* spores in milk. *Journal of the Society of Dairy Technology* 28, 80-90.

Suhren, G. (1989) Producer microorganisms. In *Enzymes of Psychrotrophs in Raw Food* ed. McKellar,R.C. pp. 3-34. Boca Raton, Florida: CRC Press Inc.

Suominen,I., Andersson,M.A., Andersson,M.C., Hallaksela,A.M., Kampfer,P., Rainey,F.A. and Salkinoja-Salonen,M. (2001) Toxic *Bacillus pumilus* from indoor air, recycled paper pulp, Norway spruce, food poisoning outbreaks and clinical samples. *Systematic and Applied Microbiology* 24, 267-276.

Sutherland,A.D. and Murdoch,R. (1994) Seasonal occurrence of psychrotrophic *Bacillus* species in raw milk, and studies on the interactions with mesophilic *Bacillus* sp. *International Journal of Food Microbiology* 21, 279-292.

Svensson,B., Ekelund,K., Ogura,H. and Christiansson,A. (2004) Characterisation of *Bacillus cereus* isolated from milk silo tanks at eight different dairy plants. *International Dairy Journal* 14, 17-27.

Taylor,J.M., Sutherland,A.D., Aidoo,K.E. and Logan,N.A. (2005) Heat-stable toxin production by strains of *Bacillus cereus, Bacillus firmus, Bacillus megaterium, Bacillus simplex* and *Bacillus licheniformis*. *FEMS Microbiology Letters* 242, 313-317.

te Giffel,M.C., Beumer,R.R., Slaghuis,B.A. and Rombouts,F.M. (1995) Occurence and characterization of (psychrotrophic) *Bacillus cereus* on farms in The Netherlands. *Netherlands Milk and Dairy Journal* 49, 125-138.

Ternström,A., Lindberg,A.M. and Molin,G. (1993) Classification of the spoilage flora of raw and pasteurized bovine milk, with special reference to *Pseudomonas* and *Bacillus*. *Journal of Applied Bacteriology* 75, 25-34.

Thiebaud,M., Dumay,E., Picart,L., Guiraud,J.P. and Cheftel,J.C. (2003) High-pressure homogenisation of raw bovine milk. Effects on fat globule size distribution and microbial inactivation. *International Dairy Journal* 13, 427-439.

Thomas,S.B. (1966) Sources, incidence and significance of psychrotrophic bacteria in milk. *Milchwissenschaft* 21, 270-275.

Thomas,S.B. and Thomas,B.F. (1973) Psychrotrophic bacteria in refrigerated bulk-collected raw milk. Part I. *Dairy Industry* 38, 11-15.

Thorsen,L., Hansen,B.M., Nielsen,K.F., Hendriksen,N.B., Phipps,R.K. and Budde,B.B. (2006) Characterization of emetic *Bacillus weihenstephanensis*, a new cereulide-producing bacterium. *Applied and Environmental Microbiology* 72, 5118-5121.

Van Netten,P., Van De Moosdijk,A., Van Hoensel,P., Mossel,D.A.A. and Perales,I. (1990) Psychrotrophic strains of *B. cereus* producing enterotoxin. *Journal of Applied Bacteriology* 69, 73-79.

Verdier-Metz,I., Michel,V., Delbès,C. and Montel,M.C. (2009) Do milking practices influence the bacterial diversity of raw milk? *Food Microbiology* 26, 305-310.

Vissers,M.M., te Giffel,M.C., Driehuis,F., De Jong,P. and Lankveld,J.M. (2007a) Minimizing the level of *Bacillus cereus* spores in farm tank milk. *Journal of Dairy Science* 90, 3286-3293.

Vissers,M.M., te Giffel,M.C., Driehuis,F., De Jong,P. and Lankveld,J.M. (2007b) Predictive modeling of *Bacillus cereus* spores in farm tank milk during grazing and housing periods. *Journal of Dairy Science* 90, 281-292.

Waes,G. (1976) Aerobic mesophilic spores in raw milk. *Milchwissenschaft* 31, 521-525.

Walstra,P. (1999) Casein sub-micelles: do they exist? *International Dairy Journal* 9, 189-192.

Walstra, P., Geurts, T. J., Noomen, A., Jellema, A. and van Boekel, M. A. J. S. (1999) *Dairy technology. Principles of milk properties and processes*. Marcel Dekker.

Wang,J.J. and Frank,J.F. (1981) Characterization of psychrotrophic bacterial contamination of commercial buttermilk. *Journal of Dairy Science* 64, 2154-2160.

Wei,J., Shah,I.M., Ghosh,S., Dworkin,J., Hoover,D.G. and Setlow,P. (2010) Superdormant spores of *Bacillus* species germinate normally with high pressure, peptidoglycan fragments and bryostatin. *Journal of Bacteriology* 192, 1455-1458.

Werner,B.G. and Hotchkiss,J.H. (2006) Continuous flow nonthermal CO_2 processing: the lethal effects of subcritical and supercritical CO_2 on total microbial populations and bacterial spores in raw milk. *Journal of Dairy Science* 89, 872-881.

Wolfson,L.M. and Sumner,S.S. (1993) Antibacterial activity of the lactoperoxidase system: a review. *Journal of Food Protection* 56, 887-892.

Woo,A.H. and Lindsay,R.C. (2006) Statistical correlation of quantitative flavor intensity assessments and Individual free fatty acid measurements for routine detection and prediction of hydrolytic rancidity off-flavors in butter. *Journal of Food Science* 48, 1761-1766.

Woods,R.G., Burger,M., Beven,C.A. and Beacham,I.R. (2001) The *aprX-lipA* operon of *Pseudomonas fluorescens* B52: a molecular analysis of metalloprotease and lipase production. *Microbiology* 147, 345-354.

Xu,S., Labuza,T.P. and Diez-Gonzalez,F. (2006) Thermal inactivation of *Bacillus anthracis* spores in cow's milk. *Applied and Environmental Microbiology* 72, 4479-4483.

Yan,L., Langlois,B.E., O'Leary,J. and Hicks,C. (1983) Effect of storage conditions of grade A raw milk on proteolysis and cheese yield. *Milchwissenschaft* 38, 715-719.

In: Raw Milk
Editors: J. Momani, et al
ISBN: 978-1-61470-641-0

Chapter 2

APPLICABILITY OF PULSED FIELD GEL ELECTROPHORESIS FOR THE IDENTIFICATION OF LIPOLYTIC AND/OR PROTEOLYTIC PSYCHROTROPHIC *PSEUDOMONAS* SPECIES IN RAW MILK

***P. D. Button*[1,2,4]*, H. Roginski*[2,5]*, H. C. Deeth*[3] *and H. M. Craven*[1]**

[1]CSIRO Food and Nutritional Sciences,
Werribee, Victoria, Australia
[2]School of Agriculture and Food Systems,
The University of Melbourne, Gilbert Chandler campus, Werribee, Victoria, Australia
[3]School of Agriculture and Food Sciences, The University of Queensland, St. Lucia, Queensland, Australia
[4] School of Applied Sciences, RMIT University, Melbourne, Victoria, Australia
[5]Department of Agriculture and Food Systems, The University of Melbourne, Parkville campus, Victoria, Australia

ABSTRACT

Many types of microorganisms are present in the milk collection environment and diversity in the raw milk microflora is typical, without dominance of a single species. The proportion of psychrotrophic bacteria in raw milk can vary widely and is associated with the level of farm hygiene. Studies in Europe have shown that typically, no more than 10% of the flora of good quality milk will be psychrotrophic with *Pseudomonas* species comprising a substantial proportion of these. *Pseudomonas fluorescens*, the most common species of the genus present in raw milk, has been involved in bacterial spikes (sudden elevations in total bacterial count) in farm bulk tank milk. Psychrotrophic *Pseudomonas* species play an important role in spoilage of UHT milk through the production of heat-stable lipases and proteases in raw milk that retain activity following UHT processing. Lipase and protease, produced by psychrotrophic *Pseudomonas* species are detected when the cell count exceeds ~10^6 cfu/mL. Prolonged refrigerated (4 °C) storage of raw milk increases the proportion of *Pseudomonas* species as do slightly higher temperatures (for example 6 °C) over a shorter period of time. This in turn increases the likelihood that they will produce heat-stable lipases and proteases. Furthermore, temperature fluctuations have been shown historically to occur in farm bulk milk, and the temperature of raw milk at the time of collection can vary widely. While less likely to occur today, both these scenarios could further compound the problem of *Pseudomonas* species proliferation in raw milk.

The aim of the present study was to investigate the use of pulsed field gel electrophoresis (PFGE) for identifying sources of lipase and/or protease producing psychrotrophic *Pseudomonas* species at various pre-processing locations, and to track the types identified through the pre-processing environment. Incubation of raw milk was also carried out to simulate possible scenarios where the raw milk may be stored on the farm and in the silo prior to UHT processing. This enabled enrichment for spoilage bacteria and studies to identify sources of microorganisms that may contribute to lipolysis and proteolysis in raw and, subsequently, UHT milk or other long life dairy products. The impact of various storage conditions on the different Pulsed Field (PF) types of importance with regard to lipase and protease production was also assessed.

INTRODUCTION

Knowledge of the microbial composition of raw milk is vital for determination of its suitability for processing into various dairy products. Some products demand use of raw milk with a low level of specific microorganisms, and unless achieved may result in quality problems in specific processed products. However, potentially of greater importance is to ensure that a given flora profile comprises a population with metabolic activities unlikely to result in a particular spoilage outcome in a particular type of product. To this end, it is imperative to quantify the psychrotrophic flora in raw milk that possesses hydrolytic enzyme capability. This is because such bacteria can be problematic for spoilage of long-life dairy products (Sorhaug and Stepaniak, 1997). This approach can also be extended for use in establishing, assessing and maintaining good agricultural practice (GAP) as outlined by the FAO (Poisot and Casey, 2007), as an important first step to complement the use of good handling practices (GHP) and good manufacturing practices (GMP) further along the supply chain. Consequently, microbiology-based methodology to reliably predict the potential for long-life dairy product spoilage could play an important role in routine quality control, establishing quality assurance programs or troubleshooting problems that occur with GAP, GHP and GMP.

Traditional identification methods for bacteria are based on phenotype, such as biochemical and growth-related (cellular and colonial morphology) characteristics. However, identification methods based on phenotypic characteristics have limitations, including lack of reproducibility and discriminatory power as well as being ineffective at providing a link between results obtained from different samples (Dogan and Boor, 2003). Hunter and Gaston (1988) state that "discrimination, reproducibility and typability" (genetic relatedness) are the most important requirements to consider when assessing typing methods. While genotypic typing is not necessary to provide identification below subspecies level (Tenover *et al.*, 1995), only genotypic methods can best satisfy these three requirements. Molecular typing methods have emerged as important techniques in determining the genetic relatedness of bacteria, especially in epidemiology studies for tracking sources of pathogenic and spoilage bacteria (Wiedmann *et al.*, 2000). A molecular-based approach to identification provides the definitive answer to the question of relatedness of bacterial isolates (Goering, 2010). Pulsed field gel electrophoresis (PFGE), developed by Schwartz and Cantor (1984), was the only one of 13 typing methods described by Maslow and Mulligan (1996), which was ranked in the top tier for all the three important criteria stated by Hunter and Gaston (1988). Consequently, it is a method widely applicable for typing of most bacteria (van Belkum *et al.*, 2007). In addition, it has been regarded as the "gold standard" for molecular typing of many bacteria for some time (Maslow and Mulligan, 1996; Goering, 2010), especially for *Pseudomonas* species, although such reports are usually based on epidemiological studies of pathogenic species, typically *P. aeruginosa*. Although PFGE is rather expensive compared with some other molecular typing methods, such as ribotyping (Wiedmann *et al.*, 2000), RAPD and PCR, and it takes a lengthy period for the entire analysis to be completed, Olive and Bean (1999) considered it the best technique for bacterial genotypic characterisation. While present research is indicating that sequencing-based techniques may be comparable to PFGE according to certain criteria, they are yet to be established and the high cost of capital equipment is hindering the potential acceptance and use of those techniques (Foley *et al.*, 2009). Consequently, PFGE still holds a solid place in the molecular typing of bacteria.

Molecular identification methods have value in the typing of spoilage bacteria to identify sources of contamination of the product (van der Vossen and Hofstra, 1996). Such an approach has been used for typing of pseudomonads contaminating milk by Dogan and Boor (2003). They used a molecular typing technique (ribotyping) to only identify *Pseudomonas* species, that were present in various areas within the dairy environment (raw milk, factory environment and pasteurised milk), and then assessed their genetic diversity and lipolytic and proteolytic potential. Jayarao and Wang (1999) investigated the diversity of *P. fluorescens* in farm bulk tank milk using phenotypic typing methods. Earlier, Ralyea *et al.* (1998) used ribotyping to track *P. fluorescens* in a dairy production system. These investigations demonstrated the suitability of molecular typing for *Pseudomonas* species within the dairy environment (bulk raw milk, pasteurised milk and various locations on the farm and in the factory) because the source of contamination was identified and the technique demonstrated a high discrimination index.

Various culture-based studies have been undertaken to investigate how widespread lipase and/or protease production is among *Pseudomonas* species isolated from raw milk. Dempster (1968) and Shelley *et al.* (1987) found that a large proportion of the lipolytic psychrotrophic flora of raw milk was *Pseudomonas* species, of which *P. fluorescens* was the species most often identified. Although less commonly found in raw milk, *P. fragi* is possibly more important in lipolytic spoilage (Shelley *et al.*, 1987). *Pseudomonas* species, particularly *P. fluorescens,* are the most frequently isolated proteolytic flora of raw milk (Ewings *et al.*, 1984; O'Connor *et al.*, 1986) and are more likely to be proteolytic than lipolytic (Wang and Jayarao, 2001). However, a high proportion of *Pseudomonas* species isolated from raw milk produce both lipases and proteases (Muir *et al.*, 1979).

Spoilage of bulk milk can originate from a small group of farms, or even a single farm. Once this poorer quality milk has been mixed with milk collected from other farms, it is impossible to identify the farm(s) contributing to the problem, unless individual sampling has been conducted at each farm. The specific sources of contaminating organisms in milk can be diverse. The on-farm contamination sources include teats and udders of the cow, particularly for *Pseudomonas* species (Desmasures *et al.*, 1997a). Improperly cleaned milking equipment has also been shown to be a significant source of psychrotrophs in farm milk (Thomas *et al.*, 1971). This and additional points along the pre-processing line may contribute as a result of biofilm development (Roberts, 1979). It can be useful to track spoilage organisms through the pre-processing chain to determine which locations need to be addressed with regard to hygiene, because this information can be used to reduce the risk of bacteria with spoilage potential being present. This leads to improved raw milk quality and consequently improved quality of processed milk and milk products.

The aim of the present study was to identify sources of lipase- and/or protease-producing psychrotrophic *Pseudomonas* species at various pre-processing locations using pulsed field gel electrophoresis (PFGE), and to track the types identified through the pre-processing environment. Established protocols, for example, with regard to selection for growth of psychrotrophs, were used to simulate possible scenarios where the raw milk is stored on the farm and in the silo prior to UHT processing. Under these expected standard growth conditions, an assessment was made of the suitability of PFGE as a technique to identify *Pseudomonas* species isolates to below species level, particularly with regard to lipase and protease production.

Materials and Methods

Chemicals, Microbiological Media and Reference Strains

Unless otherwise stated, all chemicals were purchased from Sigma-Aldrich Co. (Sydney, Australia) and were of the highest grade available. Microbiological media was purchased from Oxoid Australia Pty. Ltd. (Adelaide, Australia). *Pseudomonas fluorescens* strains ATCC948, from the American Type Culture Collection (Manassas, VA, United States), and SBW25, kindly provided by Andrew Spiers at the University of Oxford, were used for reference purposes.

Sources of Raw Milk

Five raw milk samples, each of approximately 500 mL, were collected. Three were from farms in the area around the towns of Cardinia and Bayles, just beyond the south eastern suburbs of Melbourne, Australia. Regular milk collection was on a daily basis from two of these farms (Farms 2 and 3), while from the other (Farm 1) milk was collected every second day. Farm samples were obtained by sampling directly from the bulk tank. Another sample was taken from the milk tanker used to transport the raw milk from these farms to the milk processor. The tanker had been used previously to collect milk from other farms and had not been washed before collection of milk from Farms 1, 2 and 3. The final sample was from the silo at the milk processing site. This silo contained raw milk from this single delivery of milk only, and was cleaned prior to filling with this milk.

Incubation to Achieve Spoilage Levels

A volume of 20 mL of milk was incubated statically to achieve spoilage levels under various conditions. All farm milk was incubated at 4 °C for 7 d (daily enumeration), 10 °C for 4 d (every second day enumeration) or at 4 °C for 2 d followed by 10 °C for 2 d (every second day enumeration). Silo milk was incubated at 4 °C for 4 d (daily enumeration) or 4 °C for 2 d and then 10 °C for 2 d (every second day enumeration). Milk collected from the tanker was not incubated as this does not occur in practice.

Enumeration of Aerobic Mesophiles and Enumeration, Isolation and Presumptive Identification of Psychrotrophic *Pseudomonas* Species

Enumeration was carried out using the spread plate technique, based on AS 1766.1.4 (Standards Australia, 1991). Enumeration of total aerobic mesophiles was on Plate Count Agar, incubated at 30 °C for 72 h. Enumeration of psychrotrophic *Pseudomonas* species was on Pseudomonas Agar with C-F-C supplement. Incubation for isolation of psychrotrophs was at 7 °C for 10 d (Juffs, 1972). Isolates were taken from all unincubated farm, tanker and silo milk samples and from samples of incubated milk from each farm and the silo when the total plate count had reached 10^6 cfu/mL. The relative proportions of each morphologically distinct colony type on the counted plates were recorded and one of each type selected for further investigation. Initially, this involved purification of the culture on non-selective media by subculturing into 10 mL of Nutrient Broth, incubating at 30 °C for 24 h before streaking for single colonies onto Nutrient Agar and incubating at 30 °C for 24 h. After this, Gram staining and testing for the presence of oxidase were performed. Pure isolates which were oxidase positive Gram-negative rods were considered to be psychrotrophic *Pseudomonas* species.

Screening of Bacterial Isolates for Lipase and Protease Production

An agar diffusion method based on Christen and Marshall (1984) and Craven (1993) was used to screen the isolates for lipase and protease activity respectively. Nutrient Agar plates containing either 0.1% triolein (for lipase) or 1% skim milk (for protease) were used. The Nutrient Agar plates just described were poured in equal identical layers of 10 mL each – the first was allowed to set, then the second layer was poured. Portions of the top layer of the agar were removed using a 6 mm sterile cork borer. A 10 μL aliquot of a Nutrient Broth culture incubated at 25 °C for 24 h (containing approximately 10^8 cfu/mL) was added to each well. The plates were incubated for 168 h at 4 °C with observations of zones of clearing around the wells recorded after 93 h, as well as at the end of the incubation period. Measurements were taken to the edge of the zone from the edge of the well. The largest zone size at 168 h for each test (17 mm for lipase production and 28 mm for protease production) was divided by three. This determined the designations for weak, moderate and strong producers. Isolates tested for lipase production were recorded as weak if the size of the zone, measured in the manner described above, was between one and five millimetres, moderate if between six and 11 mm and strong if between 12 and 17 mm. Isolates tested for protease production were recorded as weak if the zone was between one and nine mm, moderate if between 10 and 19 mm and strong if between 20 and 28 mm.

Extraction of Genomic DNA and Restriction Endonuclease Digestion

Cultures were inoculated into 10 mL of Tryptic Soy Broth and incubated at 25 °C for 35 h. A 1.5 mL volume of this incubated culture was centrifuged for 2 min at 10 000 *g*. The supernatant was discarded and the pellet resuspended in 1 mL of SE buffer (75 mmol l-1 NaCl, 25 mmol l-1 EDTA - pH 7.4) before identical centrifugation. Again, the supernatant was discarded and the pellet resuspended in 500 µL SE buffer. An equal volume of 2% (w/w) Sea Plaque agarose (BioWhittaker Molecular Applications; Rockland, ME, United States) was prepared in SE buffer and mixed with the cell suspension in SE buffer. Two plugs were immediately prepared in a gel mould, using 200 µL (100 µL each) of the agarose cell suspension. The plugs were allowed to set and up to 15 pieces 500-750 µM wide were sliced. All slices were immersed in 1 mL of lysis solution (500 mmol l-1 EDTA at pH 9.5, 500 µg mL-1 proteinase K and 34 mmol l-1 N-lauroylsarcosine) and incubated at 55 °C for 16 h. Following incubation, the slices were rinsed with 1 mL of SE buffer and then incubated for 15 min in SE buffer containing 1 mM phenylmethanesulphonyl fluoride (PMSF). Two further 15 min incubations were carried out with fresh 1 mM PMSF in SE buffer. After the last washing step, the 1 mmol l-1 PMSF in SE buffer was discarded and replaced with 1 mL TE buffer (10 mmol l-1 Tris base, 10 mmol l-1 EDTA - pH 7.4). Slices were stored up to one week at 4 °C in TE buffer, prior to use. Restriction endonuclease digestion was carried out with *Swa*I (New England Biolabs; Beverly, MA, United States) according to the manufacturer's instructions.

Pulsed Field Gel Electrophoresis

One percent Pulsed Field Certified Agarose (Bio-Rad Laboratories; Sydney, NSW, Australia) was prepared in 0.5X TBE buffer (Peacock and Dingman, 1967). The equipment used was the Bio-Rad CHEF-DR® II PFGE system (Bio-Rad Laboratories; Sydney, NSW, Australia). Temperature of the run buffer (0.5X TBE) was maintained at 14 °C. The initial switch time was 1.79 s and was ramped linearly to a final switch time of 1 min 33.69 s. Gradient was at 6 V/cm and the inclined angle was 120 °. Total run time was 26 h 56 min.

Band Visualisation and Data Analysis

The gel was stained in 1% ethidium bromide for 30 min followed by a brief rinse (between 30 and 60 s) in distilled water. A model TM-36 Chromato-Vue UV transilluminator (Ultra-violet Products; San Gabriel, CA, United States) was used to visualise the ethidium bromide-strained bands. Photographs were then taken with a model DC290 camera (Kodak [Australasia] Pty. Ltd.; Melbourne, VIC, Australia) operated through Kodak 1D Image Analysis Software (Eastman Kodak Company; New Haven, CT, United States), and saved as TIF images. The "Ethidium Bromide" option was selected from the "Sample type" with exposure of 4.5 s and bracket of 1.125 s. Analysis of the gel image was with the GelCompar II (Applied Maths BVBA; Sint-Martens-Latem, Belgium) software program. Dendrograms were constructed using Jeffrey's X and unweighted pair-grouping. Band matching was carried out with 1.7% position tolerance and 0% optimisation. Isolates with a similarity of at least 80% were grouped into the same PF Type.

RESULTS

Colony Counts on Fresh Raw Milk

The total counts across the five raw milk sampling sites (three farms, their milk collection tanker and the factory silo) ranged between 7.0 x 10^2 cfu/mL (Farm 2) and 9.0 x 10^3 cfu/mL (Farm 3) with a (geometric) mean of 3.2 x 10^3 cfu/mL (Table 1).

Table 1. Bacterial counts of raw milk on day of collection

Source	Number of bacteria (cfu/mL)	
	Total count	Psychrotrophic *Pseudomonas* count
Farm 1	2.5 x 10^3	~ 1.0 x 10^2 (~ 4.0)
Farm 2	7.0 x 10^2	< 1.0 x 10^2 (~ 14.3)
Farm 3	9.0 x 10^3	~ 1.7 x 10^3 (~ 18.9)
Tanker	5.0 x 10^3	~ 2.0 x 10^3 (~ 40.0)
Silo	4.0 x 10^3	~ 2.0 x 10^3 (~ 50.0)

Numbers in brackets indicate proportion of the total count as a percentage.

This mean value is close to counts obtained from the silo and the tanker. The psychrotrophic *Pseudomonas* species counts were all lower than the total counts and were lowest in milk from Farms 1 (on every second day collection) and 2 (on daily collection), which had similar counts.

The psychrotrophic *Pseudomonas* species count was similar in the samples from Farm 3 (on daily collection), the tanker and the silo, which were approximately one log higher than Farms 1 and 2. A large variation was seen in the proportion of psychrotrophic *Pseudomonas* species compared with the total plate count.

On Farm 1, these organisms comprised approximately 4% of the flora while in the silo, approximately 50% of the microbes encountered were psychrotrophic *Pseudomonas* species.

Growth of Raw Milk Microflora during Storage

At the commencement of incubation of the farm and silo milk samples, the counts of psychrotrophs were substantially lower than the total counts (Table 1).

However, after incubation for two to three days, the psychrotrophs were the predominant microflora present at all storage conditions (Figure 1).

a)

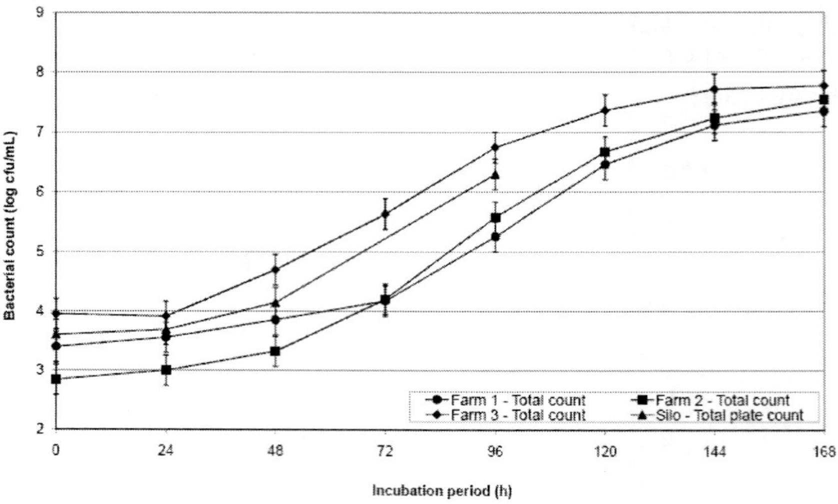

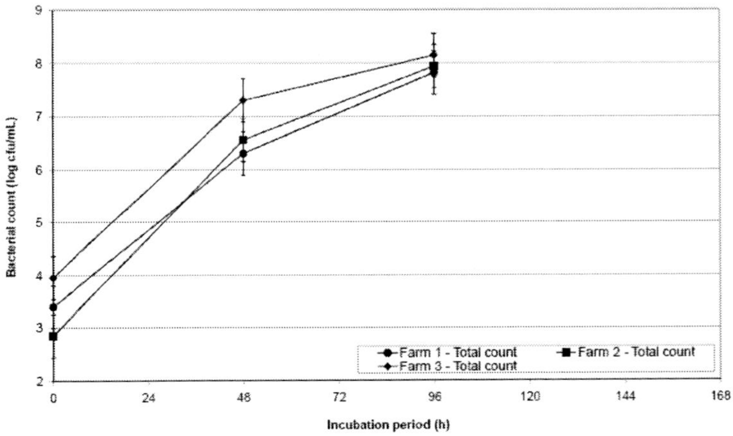

b)

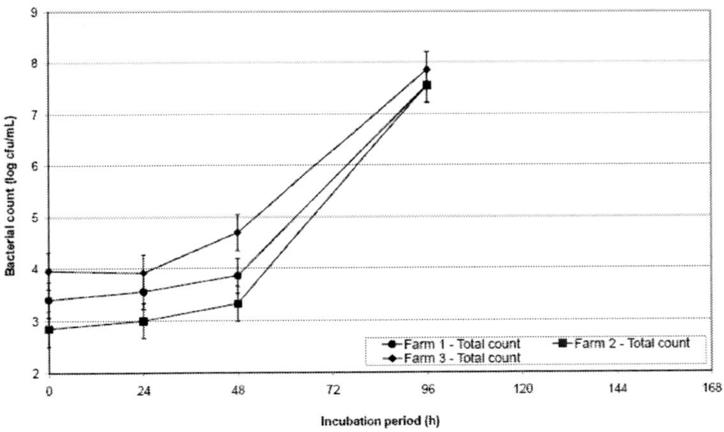

c)

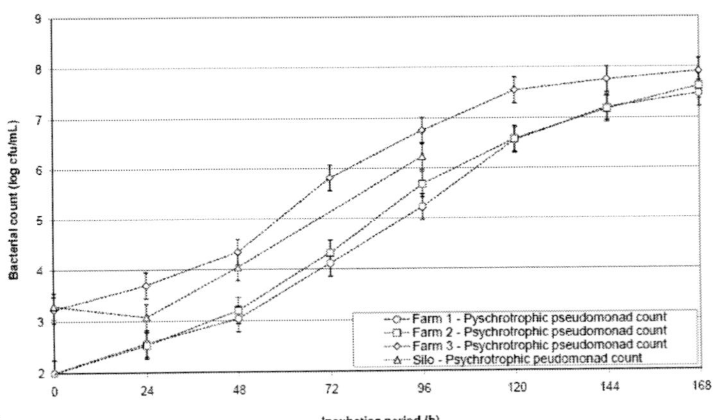

d)

Figure 1. (Continued).

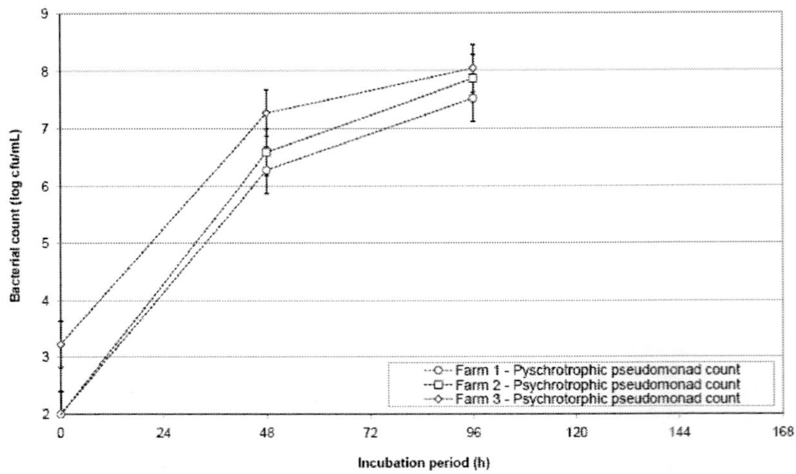

e)

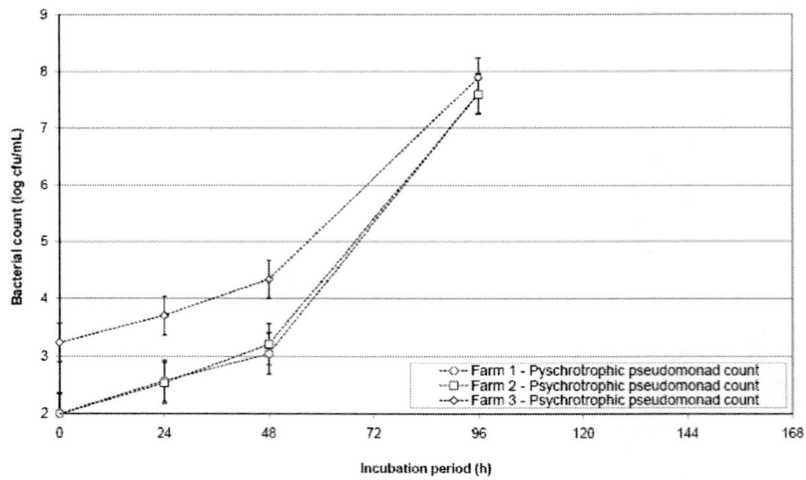

f)

Figure 1. Total count of raw milk during incubation at 4 °C (a), 10 °C (b) and 4 °C (0-2 d) followed by 10 °C (2-4 d) (c). Psychrotrophic count of raw milk during incubation at 4 °C (d), 10 °C (e) and 4 °C (0-2 d) followed by 10 °C (2-4 d) (f).

A cell count of 5 x 10^6 cfu/mL is generally regarded as the bacterial concentration where action of lipases and proteases is detectable (Law, 1979). When the milk was incubated at 4 °C (Figure 1a and 1 d), the bacterial count reached 10^6 cfu/mL in three days for milk from Farm 3, four days for the silo milk and between four and five days for milk from Farms 1 and 2. Following incubation at 10 °C, all samples contained more than 10^6 cfu/mL after two days, with the milk from Farm 3 having the highest counts of bacteria at this time (Figure 1b and 1 e). The other farm samples contained similar counts of bacteria. The fluctuating temperature of 4 °C for two days followed by 10 °C for two days led to the highest counts of bacteria in milk from Farm 3, followed by Farms 1 and 2, after two days incubation (Figure 1c and 1f). The numbers of bacteria present exceeded 10^7 cfu/mL in all samples after 4 days incubation.

Identification of Pulsed Field Types and Their Sources

A total of 45 isolates were collected from milk from three farms, their farm milk collection tanker and the silo at the factory as described above. There was much diversity in the psychrotrophic pseudomonad flora, with 39 pulsed field (PF) Types identified (Figure 2 and Table 2).

Table 2. Origin of Pulsed Field Types

Sample	Milk incubation temperature (°C)	Total number of PF Types from each location	PF Type designations
Farm 1	Not incubated	1	24
	4	2	27, 28
	10	2	24, 26
	4/10	3	4, 13, 19
Farm 2	Not incubated	2	8, 31
	4	3	7, 11, 16
	10	2	10, 21
	4/10	4	6, 7, 17, 18
Farm 3	Not incubated	2	23, 33
	4	6	3, 5, 9, 31, 38, 39
	10	4	12, 15, 20, 32
	4/10	4	2, 22, 29, 30
Tanker	Not incubated	1	36
Silo	Not incubated	2	25, 34
	4	6	1, 3, 14, 31, 35, 37

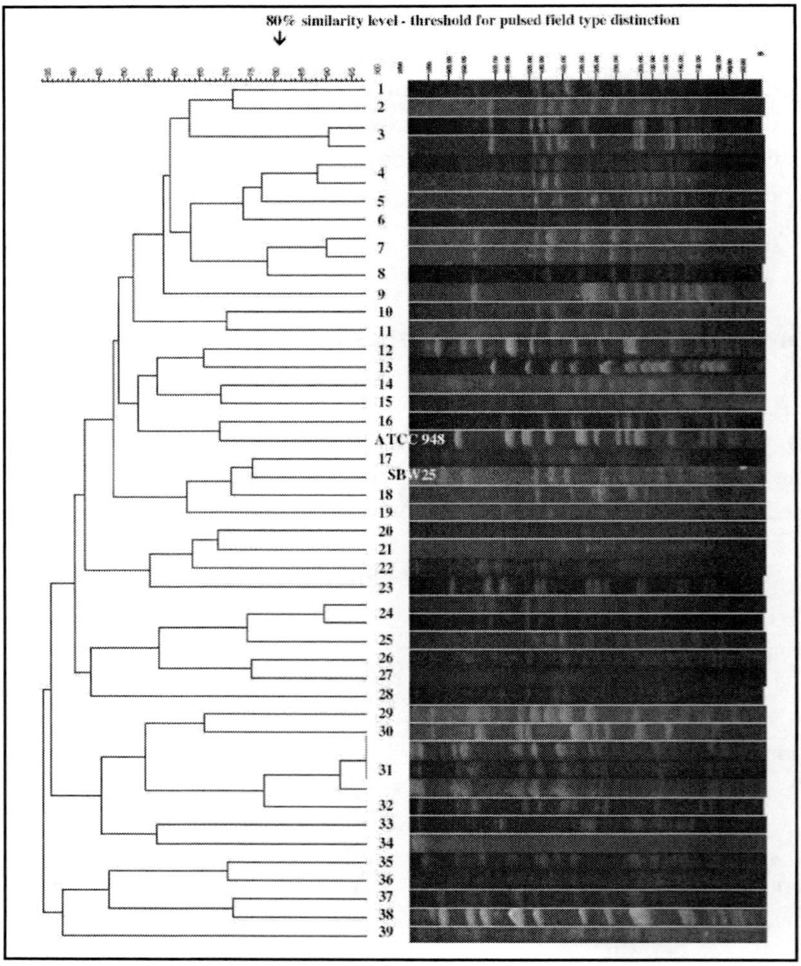

Figure 2. Dendrogram of the 39 pulsed field Types isolated from raw milk.

Of the farm samples examined, there was most diversity in the Farm 3 milk, with 16 PF Types identified from 16 isolates across all incubation conditions. Similarly, all eleven Farm 2 PF Types were from eleven isolates and all eight silo PF Types were from eight isolates. There was slightly less diversity among the Farm 1 isolates, with eight PF Types from nine isolates. The two reference isolates, *P. fluorescens* ATCC948 and SBW25, were quite distinct from most of the isolates obtained from raw milk in this study.

There were eight isolates from fresh, unincubated raw milk, all of which were from different PF Types. There was one isolate from Farm 1, two isolates from Farm 2, two isolates from Farm 3, one isolate from the tanker and two isolates from the silo.

Upon incubation at 4 °C, there was little change in the number of PF Types with milk from Farms 1 and 2 as there were two PF Types identified from Farm 1 and three PF Types identified from Farm 2. The situation was quite different for Farm 3 and the silo with six PF Types present in milk from each source. PF Type 31 was the only PF Type present in unincubated milk, that was also present in the 4 °C incubated milk. It was present prior to incubation in Farm 2 milk and then after 4 °C incubation, it was found in Farm 3 and silo milk. There was little difference in diversity of PF Types between milk incubated at 10 °C, compared to 4 °C. Both Farm 1 and Farm 2 milk contained two PF Types each while four PF Types were found in Farm 3 milk incubated at this temperature. All PF Types were unique among all of the 10 °C samples, that is, they were not found in other samples. One PF Type, 24, isolated from unincubated Farm 1 milk, was also isolated from 10 °C incubated milk from that same farm, with 89% similarity between the isolates.

PF Types obtained from farm milk incubated at 4 °C for 48 h followed by incubation at 10 °C for 48 h, were very similar across all farms, with four PF Types from each farm milk. None of the PF Types present after the 4 °C then 10 °C incubation were present in the unincubated milk, but PF Type 7 was also isolated from milk from Farm 2 after incubation at 4 °C with 92% similarity.

Sources of Moderately and Strongly Lipolytic and Proteolytic *Pseudomonas* PF Types

Table 3 presents a summary of the sources of *Pseudomonas* PF Types identified from isolates that were moderately and strongly lipolytic and proteolytic.

The isolates that were not lipolytic/proteolytic or that demonstrated weak lipolysis/proteolysis were not considered in these results because these isolates are potentially of little practical significance in the spoilage of UHT milk.

Six of the eight PF Types from the unincubated raw milk were moderate or strong lipase and/or protease producers. Four were isolated from Farm 3 (2) and silo (2) milk and the other two were from Farm 2 (1) and the tanker (1). Four of the six PF Types in this group were both lipolytic and proteolytic while two were proteolytic only.

Table 3. Origin of moderately and strongly lipolytic and proteolytic *Pseudomonas* species isolates

Sample	Milk incubation temperature (°C)	PF[1] Type designations		Total number of PF Types[2]
		Lipase	Protease	
Farm 1	Not incubated			1
	4	27	27	2
	10	26	26	2
	4/10		13	3
Farm 2	Not incubated		8	2
	4		7, 11	3
	10	21	10, 21	2
	4/10		6, 7, 17, 18	4
Farm 3	Not incubated	23, 33	23, 33	2
	4	9, 31, 39	3, 5, 9, 31, 38, 39	6
	10	12, 32	12, 15, 32	4
	4/10	30	2, 29, 30	4
Tanker	Not incubated	36	36	1
Silo	Not incubated	34	25, 34	6
	4	3, 31, 35, 37	1, 3, 14, 31, 35	1

[1] Pulsed field gel electrophoresis

[2] Total number of PF Types irrespective of whether they showed lipolytic and/or proteolytic activity.

Of the 17 PF Types isolated from the milk after incubation at 4 °C, 12 were moderate or strong lipase and/or protease producers. Again, these originated mostly from Farm 3 (6) and the silo (6). All PF Types from Farm 3 were proteolytic and three of these were also lipase producers. Five PF Types from the silo were proteolytic and four were lipolytic. Three were both lipolytic and proteolytic.

PF Types with moderate or strong lipase and protease activity were isolated from milk incubated at 10 °C from all three farms (one, two and three PF Types from Farms 1, 2 and 3 respectively). Four of the six PF Types were both lipolytic and proteolytic and two were proteolytic only. A total of eight PF Types were isolated from milk incubated at 10 °C.

Most of the PF Types isolated from milk incubated at 4 °C followed by 10 °C were not lipolytic (7 of 8 PF Types). The PF Types that were moderately and strongly proteolytic originated from Farm 1 (1), Farm 2 (4) and Farm 3 (3). Farm 3 had the only PF Type which was both lipolytic and proteolytic.

Discussion

Microbial Composition of Fresh Raw Milk

The raw milk obtained during this investigation was of good microbiological quality, with total counts ranging from 7.0 x 10^2 cfu/mL to 9.0 x 10^3 cfu/mL, depending on sampling location. From a survey of the literature by Thomas *et al.* (1971), most raw milk freshly drawn from healthy cows contains total microflora in the range from 5.0 x 10^2 to 5.0 x 10^3 cfu/mL. A later study by Senyk *et al.* (1982) reported that the total count of 86% of bulk tank milk samples was in the range 1.0 x 10^3 to 5.0 x 10^4 cfu/mL, while 92% of the psychrotrophic counts were less than 1.0 x 10^4 cfu/mL. At less than 2.0 x 10^3 cfu/mL, all milk samples in the current study were within this range. The tanker and silo total counts were also low. Some previous reports indicated that milk sampled from silos or from tankers contains higher total counts (Fryer and Halligan, 1974; Mahari and Gashe, 1990) due to contamination from the tanker or pumping and related equipment (Thomas, 1974). However, this was not observed in the present study.

In freshly drawn, good quality raw milk, psychrotrophic *Pseudomonas* species are generally present in low numbers, and are far from being the dominant microorganisms. With increasing refrigerated storage of raw milk, psychrotrophic organisms increase in proportion to dominate the flora (Cousins *et al.*, 1977). In the current investigation, between 4% and 19% of the total count in the farm samples were psychrotrophic *Pseudomonas* species Similar results have mostly been reported in the literature. However, some uncharacteristic results, by Twomey and Crawley (1968) and Chye *et al.* (2004), have also been observed. In those studies, psychrotrophic bacteria comprised less than 0.1% of the total count. The result of Chye *et al.* (2004) may reflect higher ambient temperatures in the milk collection areas. More typical values are quoted by Desmasures and Gueguen (1997), who sampled monthly from the bulk tank on four farms over two years. The mean observation was that *Pseudomonas* species accounted for between four and 23% of the total count, with two of the four farms averaging less than 5% pseudomonads. Similar low values were observed by Jaspe *et al.* (1995), with pseudomonads comprising 5% of the total count, and psychrotrophs 6%. In a smaller study by Desmasures *et al.* (1997b), there was a much higher incidence of *Pseudomonas* species, with these organisms comprising a higher proportion of the total count in winter (28%), compared to the warmer period of the year (21%).

In the present investigation, larger proportions of psychrotrophic pseudomonads were recovered from the tanker (40%) and silo (50%) samples than from the farm samples. This may reflect the growth or further addition of psychrotrophs beyond the farm. Pseudomonads are among the organisms which commonly form biofilms on food contact surfaces (Salo *et al.*, 2006) including stainless steel (Hood and Zottola, 1997). Therefore, milk contact surfaces on the farm, such as in the bulk tank, may be expected to develop biofilms. In fact, on the farm, biofilms have been known to form on milking equipment (Teixeira *et al.*, 2005). Furthermore, mixed cultures of species (as is present in raw milk) have been found to stimulate each other's capability to form biofilms (Kives *et al.*, 2005) and to resist sanitisers (Lindsay *et al.*, 2002). In the food processing environment, biofilms are of concern (Mosteller and Bishop, 1993) and with biofilms difficult to remove (Kumar and Anand, 1998), may be a source of pyschrotrophs for raw milk where suitable surfaces are available, including the tankers and silos, which are also made of stainless steel. .

Change in Cell Count of Raw Milk after Storage Simulation and the Possible Effects on Manufactured Dairy Products

During the storage simulation experiments, the proportion of psychrotrophs rose with increasing cold storage, as would be expected. Similar results, albeit over a shorter simulated storage period, were reported by Fryer and Halligan (1974). Senyk *et al.* (1988), who investigated changes in the microflora after storage at temperatures between 1.7 and 10.0 °C, found that psychrotrophs comprised a substantial portion (>70%) of the raw milk only when the incubation temperature was 7.2 or 10.0 °C. After 48 h incubation at 4.4 °C, the psychrotroph proportion was 22%, similar to the level (26%) after 24 h incubation. The lack of a prominent lag phase, observed in the present study, has also been reported by Griffiths *et al.* (1987). In that study, psychrotrophs were observed to comprise 41% of the total microflora at the commencement of the incubation period; however, after 18 h at 5 °C, they had increased their proportion to 54% while after storage at 10 °C, they made up 84%.

Until the second day of storage at 4 °C, the psychrotrophic *Pseudomonas* species count was considerably lower than the total count. However, after the second day, the total count and the psychrotrophic *Pseudomonas* species counts were similar, suggesting that after the second day, the total count was dominated by psychrotrophic *Pseudomonas* species. This is not surprising because pseudomonads have been shown to outgrow other psychrotrophic bacteria at refrigeration temperatures due to the shorter generation times (Jooste and Fischer, 1992). Within the first two days, the population of mesophilic aerobic bacteria would not grow, but remain viable. This is reflected in there being no increase in the total count during this period. However, the psychrotrophic *Pseudomonas* species count increased, from the commencement of storage in most instances, and it took approximately two days until they outnumbered the other flora.

When incubated at 10 °C, a substantial change in the time frame of the growth curve is immediately recognisable. Similar to 4 °C, the psychrotrophic *Pseudomonas* species dominated the raw milk stored at this temperature, but reached levels of 10^6 cfu/mL sooner, in approximately half the time. This is consistent with the growth pattern of psychrotrophic bacteria, which have shorter generation times as temperature increases (Greene and Jezeski, 1954), with the generation times of mesophilic and psychrotrophic bacteria being approximately equal only above 15 °C (Bester *et al.*, 1986). Psychrotrophic and mesophilic bacteria have an optimum growth temperature within the ambient range but only psychrotrophs are capable of growth at normal refrigeration temperatures (Adams and Moss, 1995). Therefore, with an increase in temperature, both psychrotrophic and mesophilic bacteria will increase in growth rate. The observation that both the total and psychrotrophic *Pseudomonas* species counts were nearly identical would suggest that psychrotrophic bacteria dominate the flora, particularly during the second half of the incubation.

The storage simulations in this study have demonstrated that the 10^6 cfu/mL spoilage threshold can be attained by psychrotrophic *Pseudomonas* species in three to five days at 4 °C or in under two days at 10 °C. Previous work has indicated that the initial cell count (Dommett and Baseby, 1986; Guinot-Thomas *et al.*, 1995a) and/or storage temperature (Griffiths *et al.*, 1987) are contributing factors to the time required to reach spoilage levels, and this was observed in the present investigation. The contribution of low quality (high microbial content) raw milk to the quality of the heat-processed product has been recognised (Griffiths *et al.*, 1988). As an example of the effect of high counts, the difference in psychrotrophic *Pseudomonas* species counts in fresh raw milk of about 3.2 log cfu/mL between Farm 1 and Farm 3 is sufficient for there to be a day difference in reaching 10^6 cfu/mL at 4 °C. Higher temperature (for example 10 °C versus 4 °C) had a similar effect in shortening the time to reach the reported 10^6 cfu/mL spoilage threshold. As every second day collection of milk from farms is not uncommon (Oz and Farnsworth, 1985) and with raw milk storage at the factory prior to processing generally 24 h or longer (Celestino *et al.*, 1996), storage of raw milk for three days prior to processing occurs (Guinot-Thomas *et al.*, 1995b). As a result, the potential of the psychrotrophic *Pseudomonas* species count in raw milk to attain the 10^6 cfu/mL spoilage threshold is clearly evident if storage temperature is not well controlled.

A change in the composition of the raw milk microflora following growth has been widely reported. Due to competition and adaptation to the prevailing conditions, some populations do not persist at their original proportions and some may disappear altogether (Lafarge *et al.*, 2004).

Pulsed Field Types in Raw Milk: Variation and Potential Impact on Manufactured Dairy Products

It was clear from the PFGE Typing results that there was much diversity among the psychrotrophic *Pseudomonas* species due to both the sample location and the incubation conditions applied to the milk, as the 45 isolates from incubated milk could be assigned into 39 PF Types. A high degree of genetic diversity has been reported in pseudomonads, based on the results of two molecular typing methods. These were (I) ribotyping, used by Dogan and Boor (2003) in a study of isolates from milk (raw from farms and pasteurised) as well as from the farm and factory environment, and (II) the random amplified polymorphic DNA (RAPD) technique, used by Martins *et al.* (2006) to characterise isolates from raw milk, from unspecified location(s).

Overall, in the present study, a greater proportion of PF Types which were moderately or strongly lipolytic or proteolytic were obtained after incubation of the milk. This demonstrates the significance of cold storage in selecting for the development of spoilage bacteria. This was particularly evident for the milk from Farm 3 and the silo, where the initial level of psychrotrophs was relatively high. Storage at 4 °C resulted in a greater proportion of bacteria with higher lipolytic and proteolytic potential than the higher incubation temperatures.

Overall, fewer PF Types demonstrated lipase production compared to protease production. An interesting observation is that strong lipase producers were also strong protease producers but strong protease producers were not always strong lipase producers. Also, without exception, PF Types devoid of proteolytic action also lacked lipolytic action. Therefore, there is a strong association between strong production of both lipase and protease or between absence of lipase and protease production. In general, the moderate and strong protease producing PF Types predominated, thereby increasing the likelihood of proteolytic spoilage in manufactured dairy products.

Percentages of *Pseudomonas* isolates from raw milk reported to produce lipase and/or protease are variable. Wang and Jayaro (2001) also observed a higher proportion of protease producing isolates (91%), compared to lipase producing isolates (46%) at an incubation temperature of 22 °C, in their samples of farm bulk tank milk in South Dakota and Minnesota, in the U.S. Conversely, in studies by Muir *et al.* (1979) and Muir and Banks (2000), lipase production was much more common, particularly among non-fluorescent *Pseudomonas* isolates. This is in contrast to the findings of Dogan and Boor (2003), who found lipase and protease production fairly equally distributed among raw milk *Pseudomonas* isolates from dairy processing plants in New York State. Clearly, microflora can differ at various locations, which necessitates specific tracking studies to investigate and rectify quality problems such as lipase and protease contamination of milk.

Importance of Raw Milk Microflora from Farm 3 and the Silo

Farm 3 appeared to be an important farm, with regard to contamination of raw milk with psychrotrophs. The highest psychrotrophic count was observed in samples from this farm compared with the others, despite the fact that milk was collected daily from this farm. Furthermore, some of the PF Types from this farm appeared to have been transferred to the silo. This is consistent with Farm 3 milk containing the highest psychrotrophic count, and therefore would contribute more psychrotrophs to the silo milk than the other two farms. However, most of the PF Types from the silo were unique to this source indicating this to be a significant source of contamination in addition to Farm 3.

After the Farm 3 and silo milk samples were incubated at 4 °C, a high proportion of strongly lipolytic and/or proteolytic PF Types were isolated. This demonstrates that if the Farm 3 or silo milk had been stored at this temperature in practice, the possibility of product contamination with heat-stable lipases and proteases from these sources would be high. It also reinforces the need to thoroughly clean refrigerated milk storage equipment to prevent the proliferation of these bacteria on surfaces which may contaminate subsequent batches of milk.

Selection of Restriction Endonuclease

Choice of restriction endonuclease is very important (McClelland *et al.*, 1987) because PFGE is a technique that requires a small number of DNA fragments to allow accurate interpretation. The widely adopted interpretation criteria of Tenover *et al.* (1995) (which are based on the number of band differences and how these band differences relate to similarity between isolates) cannot be applied easily if there are too many or too few fragments generated. It will be difficult to identify individual bands if the number of fragments are too numerous, thereby leading to a false number or position of bands. Consequently, selection of a rare-cutting restriction endonuclease can alleviate this problem (Allardet-Servent *et al.*, 1989). If there are too few bands, identifying individual bands will not be of concern, but the application of the Tenover *et al.* (1995) criteria could be equally difficult. This is because those criteria are based on the number of band differences, with a seven band difference sufficient to demonstrate unrelatedness. If a given restriction endonuclease results in fewer than ten fragments, these criteria cannot be applied reliably (Tenover *et al.*, 1995). The ideal maximum number of bands for accurate analysis of DNA restriction patterns following PFGE is between 25 (Goering, 2004) and 60 (Romling, 2004). There are, however, mathematical models available for the determination of the optimal number of bands (Mendez-Alvarez *et al.*, 1997), but these models are difficult to apply because they are too complex and cumbersome for routine use. The number of band differences need to be viewed in context of the genetic diversity of the organism (Barrett *et al.*, 2006). Indistinguishable band patterns do not mean a great deal when an organism is genetically homogeneous, but are of much importance when an organism is genetically diverse (Barrett *et al.*, 2006). When interpreting band patterns, consideration needs to be given to factors which can influence the separation and appearance of bands. For example, Barrett *et al.* (2006) explains how the presence of a plasmid, or multiple plasmids, can alter a restriction pattern enough to distinguish otherwise indistinguishable isolates as can deletions or insertions into the DNA which would result in a restriction pattern containing multiple bands of a similar size that cannot be resolved. Therefore, the criteria of Tenover *et al.* (1995) cannot be universally applied, and the information gathered from PFGE typing needs to be considered with all phenotypic and other information. Selection is made considerably easier with the complete genome sequences of many bacteria, and other microorganisms, known. Furthermore, on-line restriction digest simulators with a wide array of restriction endonucleases (Vincze *et al.*, 2003;

Bikandi *et al.*, 2004) make the task of selection relatively straight-forward. The starting point for enzyme selection is the G+C content of the genome. For example, a genome with a high G+C content will be digested best with an enzyme with an A+T recognition sequence, since such bases are rarer in the genome. Moreover, particular sequences are rare in some genomes (such as CTAG or CCG/CGG in genomes with over 45% G+C content) (McClelland *et al.*, 1987) along with length of the recognition sequence - the longer the recognition sequence, the rarer the frequency of cutting (Romling *et al.*, 2004). This means that enzymes that recognise an eight-base pair sequence are going to cleave the DNA less frequently than an enzyme that recognises a six-base pair sequence. The G+C content of *P. fluorescens* is 63.3% (Paulsen *et al.*, 2005), therefore this species is considered G+C rich. Selection of a restriction endonuclease with an 8-bp recognition sequence of only (*Pac*I, *Swa*I) or mostly (*Pme*I) A or T residues would ensure infrequent cutting. This was confirmed with the on-line restriction endonuclease digestion simulators. A further point to consider is that additional restriction endonucleases can often be useful, in order to confirm the results or to identify a difference between isolates based on increasing discrimination. Such an approach would have been useful in the present study where there was one PF Type isolated from different farms (PF Type 31). This result, although possible, would be quite unexpected, unless transfer of isolates between farms was likely. PF Type 31 could be traced to one of the two farms (Farm 3) based on its lipase and protease production (Table 2).

Pulsed Field Gel Electrophoresis for Molecular Typing of *Pseudomonas* Species

Many methods are available for molecular typing, the choice of which depends on a variety of factors. While PFGE may currently be the best available method for typing of bacteria (van Belkum *et al.*, 2007; Goering, 2010), it would not be the method of choice under all circumstances. As stated earlier, the three key criteria for a reliable molecular typing method are the typability, reproducibility and discriminatory power (Hunter and Gaston, 1988). In comparison with other molecular typing techniques, PFGE is often unsurpassed. PFGE, along with PCR, was stated as the best molecular typing technique for *Pseudomonas* species by Maslow and Mulligan (1996) who rated PFGE "excellent" for the three criteria above. In comparison, they rate PCR "excellent" for typability and reproducibility with unknown discriminatory power. Ribotyping, which has been used for molecular typing of dairy isolates of *Pseudomonas* spp (Ralyea *et al.*, 1998; Wiedmann *et al.*, 2000; Dogan and Boor, 2003) has “excellent” typability and reproducibility but only “good” discriminatory power. However, there are limitations to the PFGE technique. For example, some isolates cannot be typed due to DNA degradation during the electrophoresis run (Lukinmaa *et al.*, 2004) and comparisons between gels are difficult (Gurtler and Mayall, 2001). These difficulties together with the associated “technical demands” of the procedure and the high cost of the equipment are disadvantages in the application of PFGE (Tenover *et al.*, 1997). Technically, the long procedure is laborious (Cox and Fleet, 2003) and one of its most important disadvantages is the time, typically five days (Goering, 2004). Although set-up costs can be slightly higher than those of other molecular typing methods (Olive and Bean, 1999; Wiedmann *et al.*, 2000), the cost per isolate compares favourably with PCR and RFLP (Olive and Bean, 1999), but is considerably more expensive than ribotyping (Wiedmann *et al.*, 2000).

It would appear that PFGE is the method of choice for molecular typing of *P. fluorescens* and related raw milk pseudomonads. Therefore, an interesting piece of further work might be a global comparison of the genetic diversity of such isolates. From this, particular PF Types could be linked with phenotypes more likely to result in lipolytic and proteolytic spoilage of UHT milk.

CONCLUSION

The results demonstrate how PFGE could be utilised to identify transfer of psychrotrophic *Pseudomonas* species between locations within the pre-processing environment. Such transfer could contribute to the great genetic diversity observed among the psychrotrophic *Pseudomonas* species isolated from farm bulk tank milk and other sources within the pre-processing environment. Should this farm bulk tank milk be stored for prolonged periods at low temperature (4 °C), selection of lipolytic and proteolytic isolates of psychrotrophic *Pseudomonas* species is likely to occur. Consequently, such prolonged storage at this temperature should be avoided. From the raw milk collected in this study, it was clear that proteolytic spoilage is potentially more likely to occur than lipolytic spoilage in long-life dairy products produced from that raw milk. Accurate genetic level identification of isolates is imperative to assess molecular details of the origins and transfer patterns of lipolytic and proteolytic isolates of psychrotrophic *Pseudomonas* species. To this end, PFGE, the molecular typing technique of choice in this study, has worth for tracking of psychrotrophic *Pseudomonas* species originating in the dairy environment. In future studies of the genetic diversity of *Pseudomonas* species in raw milk, collecting multiple samples would give higher numbers of isolates, allowing a deeper insight into their genetic diversity, revealing potentially higher genetic diversity in these populations.

ACKNOWLEDGMENTS

Financial support from Dairy Australia and the Department of Primary Industries, Victoria is gratefully acknowledged. P.D.B. was the recipient of Dairy Australia funding and a Faculty Melbourne Research Scholarship from The University of Melbourne during this work.

REFERENCES

Adams, M.R. and Moss, M.O. (1995). *Food Microbiology*. Cambridge, England: Royal Society of Chemistry.

Allardet-Servent, A., Bouziges, N., Carles-Nurit, M.-J., Bourg, G., Gouby, A. and Ramuz, M. (1989). Use of low-frequency-cleavage restriction endonucleases for DNA analysis in epidemiological investigations of nosocomial bacterial infections. *Journal of Clinical Microbiology 27*: 2057-2061.

Barrett, T.J., Gerner-Smidt, P. and Swaminathan, B. (2006). Interpretation of pulsed-field gel electrophoresis patterns in foodborne disease investigations and surveillance. *Foodborne Pathogens and Disease 3* : 20-31.

Bester, B.H., Groeneveld, H.T. and Lombard, S.H. (1986). Prediction of the keeping quality of refrigerated raw milk. *South African Journal of Dairy Science 18*: 11-17.

Bikandi, J., San Millan, R., Rementeria, A. and Garaizar, J. (2004). In silico analysis of complete bacterial genomes: PCR, AFLP-PCR and endonuclease restriction. *Bioinformatics 20* : 798-799.

Celestino, E.L., Iyer, M. and Roginski, H. (1996). The effects of refrigerated storage on the quality of raw milk. *The Australian Journal of Dairy Technology 51*: 59-63.

Christen, G.L. and Marshall, R.T. (1984). Selected properties of lipase and protease of *Pseudomonas fluorescens* 27 produced in four media. *Journal of Dairy Science 67* : 1680-1687.

Chye, F.Y., Abdullah, A. and Ayob, M.K. (2004). Bacteriological quality and safety of raw milk in Malaysia. *Food Microbiology 21* : 535-541

Cousins, C.M., Sharpe, M.E. and Law, B.A. (1977). The bacteriological quality of milk for cheddar cheesemaking. *Dairy Industries International 42* : 12-17.

Cox, J.M. and Fleet, G.H. (2003). New directions in the microbiological analysis of foods. A. Hocking, G. Arnold, I. Jenson, K. Newton and P. Sutherland. (Eds.), *Foodborne Microorganisms of Public Health Significance*. (6th ed.). (pp. 103-162). Sydney, NSW, Australia: Australian Institute of Food Science and Technology.

Craven, H.M. (1993). Methods for the evaluation of psychrotrophic spoilage bacteria in pasteurised milk. PhD thesis. La Trobe University; Melbourne, VIC, Australia.

Dempster, J.F. (1968). Distribution of psychrophilic micro-organisms in different dairy environments. *Journal of Applied Bacteriology 31*: 290-301.

Desmasures, N. and Gueguen, M. (1997). Monitoring the microbiology of high quality milk by monthly sampling over 2 years. *Journal of Dairy Research 64* : 271-280.

Desmasures, N., Bazin, F. and Gueguen M. (1997b). Microbiological composition of raw milk from selected farms in the Camembert region of Normandy. *Journal of Applied Microbiology 83* : 53-58.

Desmasures, N., Opportune, W. and Gueguen, M. (1997a). *Lactococcus* spp., yeasts and *Pseudomonas* spp. on teats and udders of milking cows as potential sources of milk contamination. *International Dairy Journal 7* : 643-646.

Dogan, B. and Boor, K.J. (2003). Genetic diversity and spoilage potentials among *Pseudomonas* spp. isolated from fluid milk products and dairy processing plants. *Applied and Environmental Microbiology 69* : 130-138.

Dommett, T.W. and Baseby, L.J. (1986). Effects of storage conditions in a final factory on raw milk microbiological quality. *The Australian Journal of Dairy Technology 41* : 23-27.

Ewings, K.N., O'Connor, R.E. and Mitchell, G.E. (1984). Proteolytic microflora of refrigerated raw milk in south east Queensland. *The Australian Journal of Dairy Technology 39* : 65-68.

Foley, S.L., Lynne, A.M. and Nayak, R. (2009). Mlecular typing methodologies for microbial source tracking and epidemiological investigations of Gram-negative bacterial foodborne pathogens. *Infection, Genetics and Evolution 9* : 430-440.

Fryer, R. and Halligan, A. (1974). The silo storage of milk. *New Zealand Journal of Dairy Science and Technology 9* : 127-128.

Goering, R.V. (2004). Pulsed-field gel electrophoresis. In D.H. Persing, F.C. Tenover, J. Versalovic, Y.-W. Tang, E.R. Unger, D.A. Relman and T.J. White (Eds.), *Molecular Microbiology: Diagnostic Principles and Practice*. (pp. 185-196). Washington, D.C., United States: American Society for Microbiology.

Goering, R.V. (2010). Pulsed field gel electrophoresis: A review of application and interpretation in the molecular epidemiology of infectious disease. *Infection, Genetics and Evolution 10* : 866-875.

Greene, V.W. and Jezeski, J.J. (1954). Influence of temperature on the development of several psychrophilic bacteria of dairy origin. *Applied and. Environmental Microbiology 2* : 110-117.

Griffiths, M.W., Phillips, J.D. and Muir, D.D. (1987). Effect of low-temperature storage on the bacteriology quality of raw milk. *Food Microbiology 4* : 285-291.

Griffiths, M.W., Phillips, J.D., West, I.G. and Muir, D.D. (1988). The effect of extended low-temperature storage of raw milk on the quality of pasteurized and UHT milk. *Food Microbiology 5* : 75-87.

Guinot-Thomas, P., Al Ammoury, M. and Laurent, F. (1995a) Effects of Storage Conditions on the Composition of Raw Milk. *International Dairy Journal 5* : 211-223.

Guinot-Thomas, P., Al Ammoury, M. and Laurent, F. (1995b) Effects of Storage Conditions on the Composition of Raw Milk. *International Dairy Journal 5* : 211-223.

Gurtler, V. and Mayall, B.C. (2001). Genomic approaches to typing, taxonomy and evolution of bacterial isolates. *International Journal of Systematic and Evolutionary Microbiology 51* : 3-16.

Hood, S.K. and Zottola, E.A. (1997). Adherence to stainless steel by foodborne microorganisms during growth in model food systems. *International Journal of Food Microbiology 37* : 145-153.

Hunter, P.R. and Gaston, M.A. (1988). Numerical index of the discriminatory ability of typing systems: an application of Simpson's index of diversity. *Journal of Clinical Microbiology 26* : 2465-2466

Jayarao, B.M. and Wang, L. (1999). A study of the prevalence of Gram-negative bacteria in bulk tank milk. *Journal of Dairy Science 82* : 2620-2624.

Jooste, P.J. and Fischer, P.L. (1992). Comparative growth-rates and proteolytic activity in milk of *Flavobacterium*, *Pseudomonas* and *Acinetobacter*. *South African Journal of Dairy Science 24* : 63-67.

Juffs, H.S. (1972). Variation in psychrotroph counts obtained at the extremes of incubation prescribed by British standard 4285:1968. *The Australian Journal of Dairy Technology 27* : 27.

Kives, J., Guadarrama, D., Orgaz, B., Rivera-Sen, A., Vazquez, J. and SanJose, C. (2005). Interactions in biofilms of *Lactococcus lactis* ssp. *cremoris* and *Pseudomonas fluorescens* cultured in cold UHT milk. *Journal of Dairy Science 88* : 4165-4171.

Kumar, C.G. and Anand, S.K. (1998). Significance of microbial biofilms in food industry: a review. *International Journal of Food Microbiology 42* : 9-27.

Lafarge, V., Ogier, J.-C., Girard, V., Maladen, V., Leveau, J.-Y., Gruss, A. and Delacroix-Buchet, A. (2004). Raw cow milk bacterial population shifts attributable to refrigeration. *Applied and Environmental Microbiology 70* : 5644-5650.

Law, B.A. (1979). Reviews of the progress of Dairy Science: Enzymes of psychrotrophic bacteria and their effects in milk and milk products. *Journal of Dairy Research 46* : 573-588.

Lindsay, D., Brozel, V.S., Mostert, J.F. and von Holy, A. (2002). Differential efficacy of a chlorine dioxide-containing sanitizer against single species and binary biofilms of a dairy-associated *Bacillus cereus* and a *Pseudomonas fluorescens* isolate. *Journal of Applied Microbiology 92* : 352-361.

Lukinmaa, S., Nakari, U.-M., Eklund, M. and Siitonen, A. (2004). Application of molecular genetic methods in diagnostics and epidemiology of food-borne bacterial pathogens. *Acta Pathologica Microbiologica et Immunologica Scandinavica 112* : 908-929.

Mahari, T. and Gashe, B.A. (1990). A survey of the microflora of raw and pasteurized milk and the sources of contamination in a milk processing plant in Addis Ababa, Ethiopia. *Journal of Dairy Research 57* : 233-238.

Martins, M.L., Pinto, C.L.O., Rocha, R.B., De Araujo, E.F. and Vanetti, M.C.D. (2006). Genetic diversity of Gram-negative, proteolytic, psychrotrophic bacteria isolated from refrigerated raw milk. *International Journal of Food Microbiology 111* : 144-148.

Maslow, J. and Mulligan, M.E. (1996). Epidemiologic typing systems. *Infection Control and Hospital Epidemiology 17* : 595-604.

McClelland, M., Jones, R., Patel, Y. and Nelson, M. (1987). Restriction endonucleases for pulsed field mapping of bacterial genomes. *Nucleic Acids Research 15* : 5985-6005.

Mendez-Alvarez, S., Gaju, N. and Oliva, B. (1997). A mathematical model to determine the optimal number of fragments for comparison of bacterial chromosomic macrorestriction patterns. *Journal of Theoretical Biology 185* : 367-372.

Mosteller, T.M. and Bishop, J.R. (1993). Sanitizer efficacy against attached bacteria in a milk biofilm. *Journal of Food Protection 56* : 34-41.

Muir, D.D. and Banks, J.M. (2000). Milk and milk products. In D. Kilcast and Subramaniam, P. (Eds.). *The stability and shelf life of food*. (pp. 197-219). Cambridge, England: Woodhead.

Muir, D.D., Phillips, J.D. and Dalgleish, D.G. (1979). The lipolytic and proteolytic activity of bacteria isolated from blended raw milk. *Journal of the Society of Dairy Technology 32* : 19-23.

O'Connor, R.E., Ewings, R.E., Hayward, K.N. and O'Rourke, P.K. (1986). Numerical taxonomy of proteolytic psychrotrophs from Queensland raw milks. *Journal of Applied Bacteriology 61* : 25-38.

Olive, D.M. and Bean, P. (1999). Principles and applications of methods for DNA-based typing of microbial organisms. *Journal of Clinical Microbiology 37* : 1661-1669.

Oz, H.H. and Farnsworth, R.J. (1985). Laboratory simulation of fluctuating temperature of farm bulk tank milk. *Journal of Food Protection 48* : 303-305.

Paulsen, I.T., Press, C.M., Ravel, J., Kobayashi, D.Y., Myers, G.S., Mavrodi, D.V., DeBoy, R.T., Seshadri, R., Ren, Q., Madupu, R., Dodson, R.J., Durkin A.S., Brinkac, L.M., Daugherty, S.C., Sullivan, S.A., Rosovitz, M.J., Gwinn, M.L., Zhou, L., Schneider, D.J., Cartinhour, S.W., Nelson, W.C., Weidman, J., Watkins, K., Tran, K., Khouri, H., Pierson, E.A., Pierson III, L.S., Thomashow, L.S., Loper, J.E. (2005). Complete genome sequence of the plant commensal *Pseudomonas fluorescens* Pf-5. *Nature Biotechnology 23* : 873-878.

Peacock, A.C. and Dingman, C.W. (1967). Resolution of multiple ribonucleic acid species by polyacrylamide gel electrophoresis. *Biochemistry 6* : 1818-1827.

Poisot, A.-S. and Casey, S. (2007). Report of the FAO Internal Workshop on Good Agricultural Practices – Rome, Italy,27-29 October 2004. Rome, Italy: FAO.

Ralyea, R.D., Wiedmann, M. and Boor, K.J. (1998). Bacterial tracking in a dairy production system using phenotypic and ribotyping methods. *Journal of Food Protection 61* : 1336-1340.

Roberts, A.W. (1979). Bulk storage of milk at the dairy-its effects on product quality. *Journal of the Society of Dairy Technology 32* : 24-28.

Romling, U., Heuer, T. and Tummler, B. (2004). Bacterial genome analysis by pulsed field gel electrophoresis techniques. *Advances in Electrophoresis 7* : 353-406.

Salo, S., Ehavald, H., Raaska, L., Vokk, R. and Wirtanen, G. (2006). Microbial surveys in Estonian dairies. *Lebensmittel Wissenschaft und Technologie 39* : 460-471.

Schwartz, D.C. and Cantor, C.R. (1984). Separation of yeast chromosome-sized DNAs by pulsed field gradient gel electrophoresis. *Cell 37*: 67-75.

Senyk, G.F., Zall, R.R. and Wolff, E.T. (1982). Assessment of raw milk quality of New York state. *Dairy and Food Sanitation 2* : 318-320.

Shelley, A.W., Deeth, H.C. and MacRae, I.C. (1987). Growth of lipolytic psychrotrophic pseudomonads in raw and ultra-heat-treated milk. *Journal of Applied Bacteriology 61* : 395-400.

Sorhaug, T. and Stepaniak, L. (1997). Psychrotrophs and their enzymes in milk and dairy products: Quality aspects. *Trends in Food Science and Technology 8* : 35-41.

Teixeira, P., Lopes, Z., Azeredo, J., Oliveira, R. and Vieira, M.J. (2005). Physico-chemical surface characterization of a bacterial population isolated from a milking machine. *Food Microbiology 22* : 247-251.

Tenover, F.C., Arbeir, R.D. and Goering, R.V. (1997). How to select and interpret molecular strain typing methods for epidemiological studies of bacterial infections: a review for healthcare epidemiologists. *Infection Control and Hospital Epidemiology 18* : 426-439.

Tenover, F.C., Arbeit, R.D., Goering, R.V., Mickelsen, P.A., Murray, B.E., Persing, D.H. and Swaminathan, B. (1995). Interpreting chromosomal DNA restriction patterns produced by pulsed-field gel electrophoresis: criteria for bacterial strain typing. *Journal of Clinical Microbiology 33* : 2233-2239.

Thomas, S.B. (1974). The influence of the refrigerated farm bulk milk tank on the quality of the milk at the processing dairy. *Journal of the Society of Dairy Technology 27* : 180-189.

Thomas, S.B., Druce, R.G. and Jones, M. (1971). Influence of production conditions on the bacteriological quality of refrigerated farm bulk tank milk-A review. *Journal of Applied Bacteriology 34* : 659-677.

Twomey, A. and Crawley, W.E. (1968). The microflora of raw milk in relation to quality testing. *New Zealand Journal of Dairy Technology 2* : 120-122.

van Belkum, A., Tassios, P.T., Dijkshoorn, L., Haeggman, S., Cookson, B., Fry, N.K., Fussing, V., Green, J., Feil, E., Gerner-Smidt, P., Brisse, S. and Struelens, M. (2007). Guidelines for the validation and application of typing methods for use in bacterial epidemiology. *Clinical Microbiology and Infectious Diseases 13* : 1-46.

Van Der Vossen, J.M.B.M. and Hofstra, H. (1996). DNA based typing, identification and detection systems for food spoilage microorganisms: Development and implementation. *International Journal of Food Microbiology 33* : 35-49.

Vincze, T., Posfai, J. and Roberts, R.J. (2003). NEBcutter: a program to cleave DNA with restriction enzymes. *Nucleic Acids Research 31* : 3688-3691.

Wang, L. and Jayarao, B.M. (2001). Phenotypic and genotypic characterization of *Pseudomonas fluorescens* isolated from bulk tank milk. *Journal of Dairy Science 84* : 1421-1429.

Wiedmann, M., Weilmier, D., Dineen, S.S., Ralyea, R. and Boor, K.J. (2000). Molecular and phenotypic characterization of *Pseudomonas* spp. isolated from milk. *Applied and Environmental Microbiology 66* : 2085-2095.

In: Raw Milk
Editors: J. Momani, et al

ISBN: 978-1-61470-641-0

Chapter 3

RAW SHEEP MILK IN THE PROVINCE OF KARAK: PRODUCTION, CONSUMPTION AND HEALTH EFFECTS

Riadh AL-Tahiri
Department of Nutrition and Food Science, Faculty of Agriculture
University of Mutah, Karak, Jordan

ABSTRACT

Sheep milk characterized by its high percentage of fat (6-8%) and high protein percentage (4.2-4.8), besides it has a very pronounce organoleptic characteristics which make it ideal to produce dairy products with a very special taste and with long shelf-life (ghee, Jameed and Baladi cheese).

This article showed that a deficient milk refrigeration system in the small farm, beside the lack of sanitation during milking and handling constitute major factors in milk deterioration. Pasteurization of Baladi cheese milk and the boiling process of Baladi cheese have a great effort on improving the microbiological quality and the sensory evaluation of the final product.

Introduction

The province of Karak (south of Jordan) is characterized as being hot and dry during summer season, with maximum daily temperature of 30-40°C in this period.

Dairy production in the province of Kark is traditional products produced from raw sheeps milk. The milk has a specific chemical composition typical of extensive farming management, which includes grazing of sheep during the milking season, where natural grazing land is characterized by aromatic Jordanian plants that confer typical organoleptic feature to the milk. Diary products made by small scale home specialist producers offer individuality and variety to the consumer and are important to the rural economy in the province.

Baladi cheese, ghee and jameed (jameed is a cultured dairy product traditionally produced and consumed by Jordanian for many years. It is a free fat concentrated yogurt product and can be kept for months at ambient temperature without spoiling or losing its nutritional value) are still produced traditionally from raw sheeps milk in the Karak district. Milk is a very suitable medium for microbial growth, that is, microorganisms existing initially in it may grow and cause its deterioration. The pathogens that constitute the principal threat to the safety of the consumers are *Listeria monocytogenes, staphylococcus aureus, Salmonella* spp. and pathogenic *Escherichia coli.* There is also concern that *Brucella* spp. could be present in milk and milk product, especially those made from contaminated raw milk. Adams and Moss (1999) has pointed out that milk has long been recognized as an agent in the spread of human disease and within a few years it was appreciated that pasteurization was also providing protection against milk borne disease. Originally the main health concerns associated with milk were tuberculosis caused by *Mycobacterium bovis* and *M. tuberculosis* and Brucellosis caused by *Brucella* spp. In some parts of the world milk is still a significant source of these infections.

Staphylococcus may be isolated from the udders of cows, goats, and sheep. The animals may suffer from mastitis due to *Staphylococcus aureus.* Hobbs and Robert (1993) showed that The *Staphylococcus aureus* can be isolated from most samples of raw milk and may be found in untreated or lightly heated dairy products. Dairy cows commonly carry the *Staphylococcus aureus* on the udder and teats, and an infection, a form of bovine mastitis, can be set by the organism. This close association with the udder inevitably means that milk become infected, but *Staphylococcus aureus* can also be spread from the infected region to milking equipment, other utensils, and the hand of workers (Forsythe and Hayes 1998).

EL-Tahawy and EL-Far (2008) reported that somatic cell count (SCC) is a very important measure of the hygiene of milk, because SCC reflects the health of the udder; the principal cause of deviations from physiological levels is the inflammation of this gland that develops from infection (mastitis). In contrast to the concentration of microorganisms that cause mastitis, the SCC does not undergo any quantitative changes in the milk after it leaves the udder, which is why it is a common indicator of udder health. Also they found that monthly yield of milk per cow, milk fat, milk protein, lactose and solid not fat content decreased significantly with elevated somatic cell count.

The inflammation of the udder markedly increase the somatic cell counts in milk, leading to inferior processing characteristics and reduced acceptance of dairy products because of changes in components and properties of raw milk (Auldist and Hubble, 1998).

The negative effect of mastitis on the dairy industry include reduced shelf life of dairy products, due to undesirable sensory attributes caused mainly by lipolytic and proteolytic enzymes (Kitchen, 1981).

Dairy industries in the Middle East countries still have many problems with the quality of raw milk. This is due principally to the high temperatures recorded in the summer season, accompanied by a deficient milk refrigeration system (Mennane *et al.* 2007). Furthermore, the lack of sanitation during milking and handling constitutes an additional factor in deterioration. Microbial counts in raw milk are much higher in warm summer months than in cool winter months which have implications for the resulting dairy products (Mendia *et al.* 2007). Also Tunick *et al.* (2007) confirmed that microbes flourish in raw milk especially during warmer months.

Rosa *et al.* (*2008*) showed that, in the raw milk produced in the southern high-lands of Brazil, mean counts of 6.07 and 5.70 log cfu/ml were achieved for total viable count, which are indicative of poor hygiene conditions during milking. Also they mentioned that the high amount of total and fecal coliforms detected in raw milk is again an indication of the low hygiene in the initial steps of the cheese-manufacturing process. The detection of coliforms and pathogens in milk indicates possible contamination from the udder, milk utensils or water supply (Bonfoh *et al.,* 2003).

Fresh milk drawn from a healthy animal normally contains a low microbial load (less than 1000 ml^{-1}), but the loads may increase up to 100-fold or more once it is stored for sometime at normal temperatures (Richter *et al.*, 1992). However, keeping milk in clean containers at low temperatures immediately after the milking process may delay the increase of initial microbial loads and prevent the growth of microorganisms in milk between milking at the farm and transportation to the dairy plant (Adesiyun, 1994; Bonfoh *et al.,* 2003). Among the naturally existing micro-organisms in milk, some induce food poisoning outbreaks (Steele *et al*., 1997).

MATERIALS AND METHODS

Animal Health Status

According to the results of an agricultural census in 2008 there were more than 500000 sheep and goat in the province of Karak. The milk samples of this study were collected from sheep only, aged between 14 to 36 months. All sheep are vaccinated regularly against: *Brucella melitensis*, Anthrax bacilli, Foot and mouth disease, Sheep pox, PPR, Pneumonia, and Enterotoxaemia. All the samples were tested for the existence of Brucella, showed a negative results.

The microscopic count for 50 samples of the tested milk showed the white blood cell (leucocyte) count ranged from 150000 to 1100000 WBC/ cm^3, with a mean value of 364600 WBC/cm^3.

Processing Methods

Jameed and Ghee: *Jameed* is defatted and dehydrated yogurt made from sheep or goat's milk and sold in rock hard nuggets prepared in the spring and summer. The butterfat of the yogurt is separated by churning, accomplished by shaking the yogurt in a goat skin bag called a *shakwa*. At the moment a stainless steel tank with a very high speed agitator built in is used to separate the butter. The separated butterfat is then used to make *ghee*. It is made by heating butter to boil off the water and then filtering out the solidified proteins. Ghee is preserved by a combination of heat, which destroys enzymes and contaminating micro-organisms, and by removing water from the oil to prevent micro-organisms growing during storage. It has a long shelf life if it is stored in a cool place, using airtight, lightproof and moisture-proof containers to slow down the development of rancidity. The defatted yogurt, called *makhīd* at this point, is strained under high pressure through a cloth, concentrating it into *jameed*. The *jameed* is salted and formed by hand into small balls to be placed in the sun and dried until hard. To reconstitute the *jameed*, which is now fifty percent protein, it is soaked in water and then melted, giving its distinctive earthy flavor to the *mansaf* (Mansaf is a Jordanian dish made of lamb cooked in a sauce of Jameed and served with rice . It is the national dish of Jordan).

Baladi cheese : The unique processing method of producing Baladi cheese from raw sheep milk by cutting the curd to small cubic cuts, sprinkling the curd cuts with dry salt for two days to be solid enough to undergo the boiling process. Boiling process achieved by boiling the curd cuts in brine (16%salt w/v) for 3-5 minutes. The hot brined cheeses were then cooled and kept in a tin can, covered nearly to the top with a 16% cold salt solution and covered tightly with a tin lid and stored at ambient temperature. The cheese can be consumed directly on the second day or can be stored for 6-12 months.

Chemical Tests

Fat percentage of the samples was carried out by Gerber method (Davis 2002).

Protein percentage of the samples was carried out by Kjeldahl method using Vapodest 20 manufactured by Gerhardt- Germany.

Lactose percentage, Specific gravity, and Freezing point were carried out by using the Lactoscan 90 (Milk analyzer).

Microbiological Tests

Total bacteria counts were enumerated on plate agar (Criterion, Hardy Diagnostics, Santa Maria, CA, USA), using the pour plate technique and incubated at 30°C for 72 h (International Dairy Federation, 1991). Total surface bacteria count were enumerated on plate agar using the surface spread technique, and incubated at 30°C for 72 h. Fecal and total Coliforms group bacteria were enumerated on violet red bile agar (Criterion, Hardy Diagnostics, Santa Maria, CA, USA), after incubation for 48h at 44°C and 37°C, respectively (Rosa, *etal.* 2008). Staphylococcus was enumerated on Baird-parker agars (Hi Media Laboratories Pvt. Ltd. Mumbai, India) with egg yolk according to the method of staphylococcus count propose by Andrew (1992). Representative colonies with typical black appearance were picked, and subjected to coagulase test. Possession of the enzyme coagulase which coagulates plasma is an almost exclusive property of *Staphylococcus aureus* (Collins, *et al.* 1995).

Yeast and mould counts were enumerated according to the IDF standard method 94 (International Dairy Federation, 1980). An agar medium was employed, in which organisms other than yeast and moulds were inhibited by using chloramphenicol. After the plates were incubated at 25 °C for 5 days, the colonies were counted.

The IDF Standard Method 41 (International Dairy Federation, 1966) was used to determine the number of lipolytic organisms present in milk samples. A sugar-free nutrient agar medium of pH 7.5, containing emulsified butter fat coloured with a small quantity of the fat soluble base of Victoria blue as an indicator, was used. The hydrolysis of butter fat yields free fatty acids and changes the base into the blue dye, so that colonies of lipolytic organisms were coloured blue. The colonies were counted after incubation at 30 °C for 3 days.

Proteolytic micro-organisms were grown on plate count agar supplemented with skimmed milk reconstituted at 10% at 30 °C for 48 h. After solidification, a clear halo around the colonies was counted (Ceylan *et al.*, 2007).

Statistical Analysis

Data were analyzed using general linear model (GLM) of statistical analysis system (SAS 1998). Data were finally presented as least square means ±1Standard Error (SE) for the development of microbial number of raw sheep milk samples collected from two different sources.

RESULTS AND DISCUSSION

The author of this article and his college in the Department of Nutrition and Food Technology / Agricultural Faculty/ Mutah University has done many works on raw sheep milk in the province of Karak/ Jordan. The results of these works can be illustrated as fellows:

Raw sheep milk: The chemical composition and some physical properties of raw sheep milk from different regions of Karak district are shown in table (2).

Table 1. Classifications of cow conditions according to the number of somatic cell count

Categories of SCC	SCC range	Status of the cow
1	1000- 99 000	Normal healthy cows
2	100 000-199 000	Normal cow and required observation for mastitis
3	200 000-299 000	Cow susceptible to mastitis
4	300 000-399 000	Cow affected with subclinical mastitis
5	> 400 000	Cow suffering from mastitis

According to EL-Tahawy and EL-Far (2010).

Table 2. The result of composition, Physical properties, and pH of raw sheep milk collected from three Regions of Karak

Treatment	Mean Value Reg.1 Reg.2 Reg.3	Least significant difference (LSD value) $LSD=(S)^2$ S = standard deviation
Fat%	6.883 7.033 6.867	0.1151
Protein%	4.483 4.550 4.483	0.08136
Lactose%	4.767 4.767 4.767	0.04068
Specific Gravity	1.033 1.034 1.033	0.01286
Freezing Point	-0.539 -0.5398 -0.5392	0.01286
pH	6.733 6.700 6.717	0.03151

Number of samples for each Region (Reg.) = 6.

The statistical results according to DMRT at 0.005 population show that Fat% at Region 2 was significantly higher than Regions 1, and 3. The statistical program. Michigan Statistics System (MSTAT) (Russel D Freed and Scott P Eisensmith, Crop and Soil Department, Michigan state University, USA). (According to AL-Tahiri *et al.* 2008).

AL-Tahiri (2010) reported on his microbiological study of raw sheep milk produced in Karak, at two different places. The first place is a breeding station for Awassi sheep species, which has the facilities of milk refrigeration. Baladi cheese is the main product of the station. The second place is cooperative dairy plant collecting milk from the farmers to produce ghee and jameed. Most farmers have no facilities for milk refrigeration. The statistical analysis showed that there is a significant difference ($P<0.001$) in bacterial load between the samples collected from the breeding station and the samples collected from the farmers. The results for total colony count by the pouring technique, total colony count by the surface technique, total coliform, fecal coliform , yeast and mould , staphylococcus bacteria , lipolytic bacteria, and proteolytic bacteria for the milk samples collected from the breeding station showed no significant difference through the month's period ($P> 0.001$) with a mean value of 5.3, 6.4, 5.16, 3.5, 4.65, 4.40, 5.44, and 3.2 log cfu/ml respectively, while the total colony count by pouring technique for the samples collected from the farmers was significantly changed ($P<0.001$) through the month's period from log 7.4 cfu/ml at March. To log 7.95 at April to log 8.57 at May, and for the total colony count by surface technique changed from log 8.4 to log 8.9 to log 9.6, and for total coliform changed from log 6.6 to log 6.8 to log 6.96, and for fecal coliform changed from log 4.6 to log 4.95 to log 5.2, and for yeast and mould changed from log 5.78 to log 6.0 to log 6.2 and for staphylococcus bacteria changed from log 3.16 to log 3.41 to log 3.61, and for lipolytic changed from log 5.34 to log 5.59 to log 5.64, and for proteolytic changed from log 3.47 to log 3.77 to log 3.96 cfu/ml. These results are shown in Table (3) and Table (4).

It's evident that milk collected from the breeding station is less contaminated than the milk collected from the farmers, (except for staphylococcus bacteria). It is worth noting that this station is a well-organized farm whose milking is done mechanically, and storage of milk done at low temperature (4°C). Consequently it is characterized by high sanitation levels compared to the farmers, where mechanical milking is lacking and is also accompanied by a deficient milk refrigeration system and more contaminating sources may also exist. The heavy contamination of the raw milk in the farmer's milk can be explained by the earlier observations concerning milking conditions. AlMahadin (2007) showed that most tested samples of farmer's milk and milk products in Jordan including Karak area had a cfu/ml or g for bacteria and fungi beyond the maximum level of acceptance in dairy products. Most of the examined milk samples from the farmers would be qualified as poor quality raw milks according to French or American standards. The maximal bacterial load tolerated by both regulations is respectively, $5X10^5$ cfu/ml, and 3X105 cfu/ml (Oliver *et al.,* 1999). The maximum value in the farmers samples has been found in the period of relatively high ambient temperature (May), whereas the minimum value has found in the cold period (March).

Table 3. Least Square Mean (LSMean) ± Standard Error(SE) for all microbial groups in milk samples (log cfu/ml) collected from two sources: Sheep Breeding station (BS) in Karak and the collecting centre of farmers milk (FM) at dairy plant in Karak during March

source	TCC/P	TCC/S	Total coliform	Total fecal coliform	Yeast and mould	Staph.	Lipolytic bacteria	Proteolytic bacteria
BS	5.29^{a} ±0.033	6.39^{a} ±0.0833	5.15^{a} ±0.024	3.48^{a} ±0.0138	4.60^{a} ±0.014	4.55^{a} ±0.085	5.47^{a} ± 0.010330	3.22^{a} ±0.0213
FM	7.37^{b} ±0.033	8.42^{b} ±0.0833	6.60^{b} ±0.024	4.64^{b} ±0.0138	5.78^{b} ±0.014	3.16^{b} ±0.085	5.34^{b} ±0.01033	3.47^{b} ±0.0213

Results are reported as least mean values for duplicate samples from 5 replicates analysed in March. Means within column by different superscript are significantly different ($P<0.001$). TCC /P ,Total colony count /pour plate technique ; TCC /S, Total colony count / surface spread technique ; Staph, Staphylococcus bacteria. (According to AL-Tahiri, 2010).

Table 4. Least Square Mean (LSMean) ± Standard Error(SE) for all microbial groups in milk samples (log cfu/ml) collected from two sources: Sheep Breeding station in Karak(BS) and the collecting centre of farmers milk (FM) at dairy plant in Karak during May

source	TCC/P	TCC/S	Total coliform	Total fecal coliform	Yeast and mould	Staph	Lipolytic bacteria	Proteolytic bacteria
MBS	5.312^{a} ± 0.033	6.34^{a} ± 0.0833	5.171^{a} ± 0.024	3.484^{a} ± 0.0138	4.68^{a} ± 0.014	4.193^{a} ± 0.085	5.42^{a} ± 0.01033	3.197^{a} ±0.0213
MCC	8.566^{b} ±0.033	9.64^{b} ±0.0833	6.963^{b} ±0.024	5.204^{b} ±0.0138	6.20^{b} ±0.014	3.61^{b} ±0.085	5.638^{b} ±0.01033	3.961^{b} ±0.0213

Results are reported as least mean values for duplicate samples from 5 replicates analysed in May. Means within column by different superscript are significantly different ($P<0.001$). TCC /P :Total colony count /pour plate technique; TCC/S: Total colony count / surface spread technique; Staph: Staphylococcus bacteria. (According to AL-Tahiri, 2010).

The farmers do not have refrigeration equipment and in some cases have to hold their production at ambient temperature before its transport to the factory counter. If this did not affect directly and significantly the final mixture during winter, it's bound to lead to a high load in the summer. Other points should be noted such as the storage of the evening milk at room temperatures and it's mixing with the milk of the following day. Cassoli *et al.* (2007) showed that 43.9% of raw milk samples collected from Brazilian farmers has a high level of bacterial contamination (> 1000 000 cfu/ml). Chye *et al.* (2004) indicated that the mean counts per ml for total plate count was $12x10^6$ in raw milk samples collected from 360 dairy farms in Malaysia, also they confirmed that approximately 90% of the samples were contaminated by coliform bacteria and 65% were *Escherichia coli* positive, with mean counts ranged from 10^3 to 10^4 cfu per ml, *Staphylococcus aureus* were isolated from more than 60% of the samples and the mean count per ml was $12x10^3$.and these mostly showed a similarity to the results of Al-Tahiri (2010).

The use of potable chlorinated water and caustic soda as detergents for cleaning the equipment, in addition to the use of hot water for equipment sterilization in the breeding station are the reasons for the low bacterial load compared to the farmer's milk. In order to reduce contamination of milk, utensils used for milking should be rinsed, cleaned using detergent and disinfected immediately after use (Dodd and Phipps, 1994; Food and Agriculture Organization and World Health Organization, 1997). Milk can be easily contaminated by infected food handlers who practice poor personal hygiene or by water containing human discharge. Coliform and fecal coliform bacteria are often used as indicator microorganisms, and a risk that other enteric pathogens may be presented in the sample. Compared to other developing countries, the data of coliform count revealed in the study of Karak raw sheep milk for samples collected from the breeding station ($1.1x10^5$-$1.8x10^5$ cfu/ml) are, in general, equal or less than those observed for raw milk in Malaysia ($1.7x10^5$ cfu/ml) (Chye *et al.* 2004) and the loads observed in Morocco ($2.1x10^3$ -$21.7x10^5$ cfu/ml) (Afif *et al.* 2008), but for the samples collected from the farmers ($2.1x10^6$- $9.9x10^6$ cfu/mL) they are in general, higher or similar to these observed for raw milk in Malaysia and Morocco. Srairi *et al.* (2009) showed that overall milk hygienic quality in Morocco was poor (Aerobic plate count and coliforms counts were 100 fold international norms), due essentially to a lack of hygiene and inadequate milking conditions. Chye *et al.* (2004) indicated that the existence of coliform bacteria may not necessarily indicate a direct fecal contamination of milk, but more precisely as an indicator of poor hygiene and sanitary practices during milking and further handling.

Milk samples from the breeding station showed a significantly higher Staphylococcus count than farmers milk .This, mostly due to the use of milking machine in this station.. Milking machine may account for mastitis cases. Incorrect vacuum or pulsator settings or worn teat cup liners all can enhance the role of the milking machine in contributing to intramammary infection. *Staphylococcus aureus* is widely recognized as a major causative agent of clinical and subclinical mastitis in dairy cattle (Chye *et al.* 2004).

AL-Tahiri *et al.* (2008) showed that the results of their work for raw sheep milk in three regions of the Karak district indicates that mainly *L.monocytogenes* and to lesser degree *L.ivannovii* and *L.innocua* were found in raw milk and that this mostly agreed with the results of detecting the Listeria in raw milk in some of the Syrian dairy products by Abou-Younes *et al.*(2005). Baladi Cheese: Haddadin and AL-Tahiri (2010) have done a work on producing Baladi cheese from raw sheep milk made according to traditional and modified methods. The main changes in the modified method were: 1- Pasteurization (73°C for 16s) of the milk, 2- Storing the cheese at low temperature (8°C for 48h)) during the cheese sprinkling with dry salt, 3- Achieving 86°C at the centre of the cheese pieces through the final cheese boiling stage. Generally, chemical parameters of the cheese were not affected by the modification, while the microbiological quality of the cheese improved significantly, Coliform, Fecal coliform and Staphylococcus destroyed completely. The final cheese produced by the modified method had a homogenous structure with firmly packed curd, not easy broken, with no pinhole gas opening, where is the cheese manufactured with unpasteurized sheep milk were described as a sponge-like structure, not firmly packed curds and contain a high number of pinhole-gas opening caused a major disruption of the matrix. Baladi cheese may be classified as a medium moisture cheese (Kosikowski 1977), high fat cheese (22-26%), with 24-28% protein and high salt (5-10%) content. The final products yield was about 18-20%, with a final pH ranging from 5.5-6.5.

CONCLUSION

Results clearly indicated that the microbiological quality of raw sheep milk produced by the farmers in Karak was lower than required by international standards. Milk refrigeration and transportation are the main factors that influence directly the microbial quality of milk. These are the key areas that remain of relevance to milk hygiene intervention. This may require some measures such as constructing small cooling milk tanks near the farmers to cool the milk directly after milking, besides, improving water quality, hygiene practices with respect to equipment cleaning and disinfection to ameliorate the quality of raw milk in the province.

Since pathogen organisms can be found in raw milk, and can survive the manufacturing process of most unpasteurized dairy products, and some can grow at refrigeration temperature such as Listeria, it would be advisable to impose a compulsory regulation for never to use raw milk for drinking purposes or for processing any dairy products unless effective pasteurization or some other effective method of heat treatment has been applied.

Producing Baladi cheese by the modified method improves the microbiological and organoleptic quality of the cheese without significantly changing its typical properties.

The production of Jameed depends on removing all the fat from the sheep-milk yoghurt before drying. Removing the fat from the Jameed will prevent the development of rancid flavors which can badly affect the final product. This product can be considered as a safe product due to the fact, that all pathogenic microbes will be killed during the severe heat treatment of cooking during the preparation of Mansaf dish.

REFERENCES

Abou-Younes A, Slik S and Abou-Ghorrah S (2005) Detecting the Listeria in raw milk and some of the Syrian dairy products. *Journal of Damascus University for Agricultural science* 21 477- 487.

Adams, M.R. and Moss, M.O.(1999)Food Microbiology. Third Edition. The Royal Society of Chemistry.

Adesiyun, A. A. (1994). Bacteriological quality and associated public health risk of pre-processed bovine milk in Trinidad. *International journal of Food Microbiology*, 21: 253-261.

Afif, A., Faid, M.. Chigr, F., and Najimi, M. (2008). Survey of the microbiological quality of the raw cow milk in the Tadla area of Morocco. *International Journal* of *Dairy Technology*, 61: 340-346.

AlMahadin, S. M. (2007). Isolation and identification of *Listeria* from dairy products in Jordan using conventional methods and molecular technique. M.S. Thesis. Al-Balqa' Applied University, Jordan.

AL-Tahiri, R., Sharaf, O., and Rewashdeh, A. (2008). A study of the occurrence of *Listeria* species in raw sheep milk. *International Journal* of *Dairy Technology*, 61: 347-351.

AL-Tahiri, R. (2010). Study on the microbiological quality of raw sheep milk in the Karak area of Jordan. *Mu'tah Lil-Buhuth wad-Dirasat,* 25: 21-36.

Andrews, A., (1992). Manual of Food Quality Control 4. Rev. 1- *Microbiological analysis*. Published by Food and Agriculture Organization of the United Nation.

Auldist, M. J. and Hubble, I. B. (1998). Effect of mastitis on raw milk and dairy products. *Australian Journal of Dairy Technology*, 53: 25-36.

Bonfoh, B., Wasem, A., Traore A. N., Fane, A., Spillmann, H., Simbe, C. F., Alfaroukh, I. O., Nicolet, J., Farah, Z. and Zinsstag, J. (2003) Microbiological quality of cows milk taken at different intervals from the udder to the selling point in Bamako (Mali). *Food Control,* 14: 495-500.

Cassoli, L. D., Machado, P. F.,. Rodrigues, C. D. O. and Coldebella, A. (2007) . Correlation study between standard plate count and flow cytometry for determination of raw milk total bacterial count. International Journal of Dairy Technology, 60: 44-48.

Ceylan, Z. G., Caglar, A. and Cakmakci, S. (2007) . Some physicochemical, microbiological, and sensory properties of tulum cheese produced from ewe's milk via a modified method. *International Journal of Dairy Technology*, 60: 191-197.

Collins, C. H., Lyne, P.M. and Grange, J. M. (1995). *Microbiological Methods*. 7th edn., Butterworth-Heinemann Ltd.

Chye, F. Y., Abdullah, A. and Ayob, M. K. (2004). Bacteriological quality and safety of raw milk in Malaysia. *Food Microbiology* , 21: 535-541.

Davis J G (2002) *Milk Testing*, India: Agrobios. Hobbs B C and Roberts D (1993) *Food Poisoning and Food Hygiene*, 6 th edn. Edward Arnold.

Dodds, F. H. and Phipps, R. H. (1994) . Dairy management and health. In: *Milk production in developing countries*, A. J. Smith, (Eds) Centre for Tropical Veterinary Medicine, University of Edinburgh, Edinburgh, Scotland, UK.(PP: 258-271).

EL-Tahawy, A. S., and EL, Far, A. H. (2010). Influences of somatic cell count on milk composition and dairy farm profitability. *International Journal of Dairy Technology*, 63: 463-469.

Food and Agriculture Organization and World Health Organization, (1997). General requirements (Food Hygiene). Codex Alimentarius, Vol. 1B (Suppl). Food and Agriculture Organization, Rome.

Forsythe, S.J. and Hayes, P.R.(1998).Food Hygiene, Microbiology and HACCP. Third Edition. A Chapman and Hall Food Science Book.

Haddadin, J. and AL-Tahiri, R. (2010). Physicochemical, microbiological and sensory characteristics of Baladi, a traditional Jordanian cheese made from raw sheep milk. *Italian Journal of Food Science,* 22: 60-69.

Hobbs, B. C. and Roberts, D. (1993) . *Food poisoning and food hygiene.* 6th end., Edward Arnold.

International Dairy Federation. 1966. *Standard Method for the count of lipolytic organisms*. IDF stand. 41. Brussels: Int. Dairy Fed.

International Dairy Federation 1980. *Milk and Milk Products: Enumeration of Yeast and Moulds*. Standard 94. Int. Dairy Fed., Brussels, Belgium.

International Dairy Federation 1991. Milk and Milk Products: *Enumeration of Microorganisms*. Standard 100 B. Int. Dairy Fed., Brussels, Belgium.

Kitchen, B. J., 1981. Review of the progress of dairy science: bovine mastitis: milk compositional changes and related diagnostic tests. *Journal of Dairy Research,* 48: 167-188.

Kosikowski, V. F. (1977). *Cheese and Fermented Foods*, 2nd ed. P. 570. Edwards Brothers, Inc.Ann Arbor, MI, USA.

Mendia, C., Ibanez, F. C., Torre, P. and Barcina, Y. (2007). Influence of the season on proteolysis and sensory characteristics of Idiazabal cheese. *Journal of Dairy Science,* 83 : 1899-1904.

Mennane, Z., Ouhssine, M., Khedid, K. and Yachioui, E.L. M. (2007). Hygienic quality of raw cow's milk feeding from domestic waste in two regions in Morocco. *International Journal of Agriculture and Biology*, 9: 46-48.

Oliver, S. P., Lewis, M. J., Gillespie, B. E., Ivey, S. J., Coleman, L. H., Almeida, R. A., Fang, W. and Lamar, K. (1999) . Evaluation of post milking tests disinfectant containing a phenolic combination for the prevention of mastitis in lactating dairy cows. *Journal of Food protection*, 62: 1354-1357.

Richter, R. L., Ledford, R. A. and Murphy, S.C. (1992). Milk and milk products. In:V. Vanderzant , and D. F. Splittstoesser, *Compendium of methods for the microbiological examination of foods,*(pp: 837-838). Washington, DC: American Public Health Association. D.

Rosa, T. D.,. Wassermann, G. E., Volken De Souza, C. F., Caron, D., Carlini, C. R. and Ayub, M. A. Z. (2008) . Microbiological and physicochemical characteristics and aminopeptidase activities during ripening of Serrano cheese. *International Journal of Dairy Technology*, 61: 70-79.

SAS User's Guide: Statistics. Version 7.00 Edition. 1998 SAS. Inst., Inc., Cary, NC.

Srairi, M. T., Benhoda, H.., Kuper, M. and Le Gal, P. Y. (2009) . Effect of cattle management practices on raw milk quality of farms operating in two-stage dairy chain. *Tropical Animal Health Production*, 41: 259-272.

Steele, M. L., McNab, W. B., Poppe, C., Griffiths, M. W., Chen, S., Degrandis,S. A., Fruhner, L. C., Larkin, C. A., Lynch J. A. and Odumeru, J. A. (1997). Survey of Ontario bulk tank milk for foodborne pathogens. *Journal of Food* Protection , 60: 1341-1346.

Tunick, M. H., Van Hekken, D. L., Call, J., Molina-Corral, F. J. and Gardea, A., A. (2007). Queso Chihuahua : Effect of seasonality of cheese milk on rheology. *International Journal of Dairy Technology,* 60 : 13-21.

In: Raw Milk
Editors: J. Momani, et al

ISBN: 978-1-61470-641-0

Chapter 4

RAW MILK: PRODUCTION, CONSUMPTION AND HEALTH BENEFITS

Marcelo A. Ferraz[1], Claudio Antonio Versiani Paiva[2], Marcelo R. Souza[3] and Mônica M. O. P. Cerqueira[3]

[1] Food Engineering, M.Sc. Animal Science, Brazil
[2] Secretary of Agriculture/Federal District, Brazil
[3] Professor at Veterinary School/Universidade Federal de Minas Gerais state, Brazil

ABSTRACT

The milk production has been growing around the world, but the biggest growth is in South and North America (Brazil and USA) and Asia (India and China). World cow's milk production in 2008 stood at over 578 million tones, with the top ten producing countries representing about 55.4% of production. Countries with advantage on land and animal feed will be a differential of productivity, such as India, China and Brazil. The consumption has grown following the increase in population and income. The countries from North America and Oceania are the biggest consumer, but don't consume the needs, which is about 200 liters per capita per year (WHO). The lowest consume is observed in countries from Asia and Africa, but just in this countries are observed the biggest growth in income. The quantity of milk's ingestion must be considered, since the vitamins and supplements are necessary to bones, muscles and immune system. Health benefits of milk included good bone health, robust skin, good immune system, prevention of illnesses such as hypertension, dental decay, dehydration, respiratory problems, obesity, osteoporosis and even some forms of cancer. The beneficial health nutrients obtained from milk are mandatory for human body and help in prevention of chronic ailments. Keeping away severe illnesses and harmful factors can be done through increasing milk consumption.

Keywords: Milk: Production, Consumption and Health Benefits.

1. Production

The agribusiness of milk is one of the most important activities for Brazilian economy generating an income of about 15 billion of Reais (Zoccal and Carneiro, 2008). Milk is produced in all the country and differences among the systems, technologies, herds, and management are related to the efficiency and productivity. It has been estimated that more than three million of people are direct or indirectly employed in this activity (Alvim et al., 2002).

Brazil is one of the greatest milk producers in the world and it would be reached the fifth place in 2010, without considering the whole European Union as only one producer in this ranking (FAO, 2011). According to data published by USDA (2010), Brazil projected milk production for 2011 can be higher than China's production, reaching the production of Former Soviet Union. In this situation, Brazil would reach the fourth place including also the European Union-27 in 2011 (Table 1).

According to FAO (2008), Brazil produced about 6.3 billions of liters of milk in the beginning of the 60th. In 2007, almost 26.9 billion of liters were produced (Figure 1).

The growth average of milk production in Brazil from 1961 to 2007 was 4.0% a year, approximately. In the 60th and 80th decades, the growth was not related only to increase of number of milked cows but also to higher productivity. In the 70th, it was observed an increase of milked cows and a decrease of productivity.

On the contrary, in the 90th, the number of milked cows decreased, but it was noted increase of productivity. More recent data show that the milk production growth from 2000 to 2007 was due to higher number of milked cows and best productivity (Table 2) (EMBRAPA, 2008).

Table 1. Cows milk: summary for selected countries (1,000 Metric Tons)

COWS MILK: SUMMARY FOR SELECTED COUNTRIES 1,000 Metric Tons						
Fluid Milk Production	2006	2007	2008	2009	(p) 2010	(f) 2011
North America						
Canada	8,041	8,212	8,270	8,280	8,350	8,350
Mexico	10,051	10,657	10,907	10,866	11,176	11,330
United States	82,455	84,211	86,174	85,874	87,450	88,690
Sub-total	100,547	103,080	105,351	105,020	106,976	108,370
South America						
Argentina	10,200	9,550	10,010	10,350	10,600	11,070
Brazil	25,230	26,750	27,820	28,795	29,948	30,846
Sub-total	35,430	36,300	37,830	39,145	40,548	41,916
European Union - 27 1/	132,206	132,604	133,848	133,700	134,200	134,700
Former Soviet Union						
Russia	31,100	32,200	32,500	32,600	31,740	31,400
Ukraine	12,890	11,997	11,524	11,370	10,950	10,570
Sub-total	43,990	44,197	44,024	43,970	42,690	41,970
South Asia						
India	41,000	42,890	44,500	48,160	50,300	52,500
Asia						
China	31,934	35,252	34,300	28,445	29,100	30,500
Japan	8,137	8,007	7,982	7,910	7,790	7,800
Sub-total	40,071	43,259	42,282	36,355	36,890	38,300
Oceania						
Australia 2/	10,395	9,870	9,500	9,326	9,400	9,700
New Zealand 3/	15,200	15,640	15,141	17,397	16,897	18,642
Sub-total	25,595	25,510	24,641	26,723	26,297	28,342
TOTAL SELECTED COUNTRIES	418,839	427,840	432,476	433,073	437,901	446,098

Source: Counselor and attache reports, official statistics, and results of office research/.

Notes:

(p) Preliminary.

(f) Forecast.

(1) Based on deliveries

(2) Year ending June 30 for the period 2006-2008.

(3) Year ending May 31 for the period 2006-2008.

Source: USDA (2010).

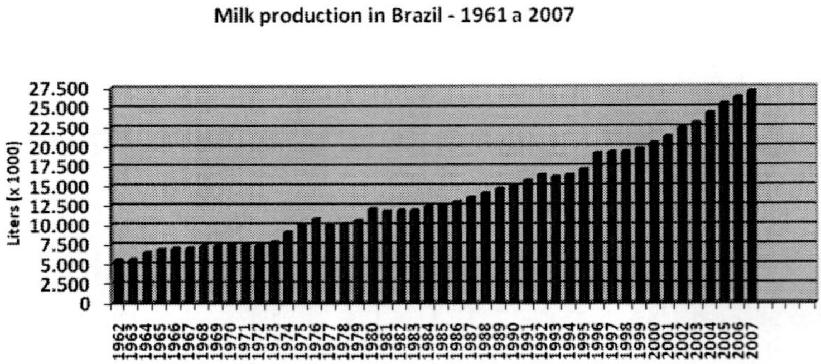

Source: FAO (2008).

Figure 1. Evolution of milk production in Brazil from 1960 to 2007.

Table 2. Factors related to milk production growth in Brazil: evolution of milked cows and production by milked cow

Period	Evolution	
	Milked cows (%)	Production by milked cow (%)
60	25%	11
70	57%	-9
80	13%	7
90	-9%	44
2000 to 2007	16%	14

Source: EMBRAPA (2008).

Gomes (2002) related that in the last decades the improvement of productivity was very significant and correlated to increase of Brazilian milk production. During the 90th, new regions that did not have tradition in milk production, such as Goiás, Triângulo Mineiro and Alto Paranaíba, began to show a better participation in this activity. In the beginning of 2000th, it was noted a new movement of expansion in states as Rondônia, Mato Grosso, and Mato Grosso do Sul.

Although the production in the new areas and the continuous growth of milk production in Brazil in the last decades, some economic indicators, such as productivity per cow, are still considered very low when compared with other countries (Table 3) (Embrapa, 2008).

Table 3. Productivity of milk per cow in several countries of the World

Country	Productivity (Kg/cow/year)
United States of America	9.219
Denmark	8.288
Canada	7.960
Netherlands	7.450
Japan	7.434
United Kingdom	7.189
Germany	6.923
France	6.240
Italy	6.064
Mexico	5.962
Australia	5.131
Argentina	4.773
Polon	4.327
New Zeland	3.817
Ukraine	3.675
Russian Federation	3.399
China	3.109
Turkey	2.529
Ira	1.500
Brazil	1.224
Paquistan	1.200
India	1.109

Source: EMBRAPA (2008).

According to Martins (2005), in Brazil, it is possible to find different systems of production in the same region that varied since intensive production to production on grazing systems that are economically and technically viable in all the country. This author describes that Brazil has one of the lowest costs of milk production together with Argentina, Uruguai, Australia, New Zealand, Chile, and India.

Considering the geographic regions of milk production in Brazil, Southeast, South and Mid-west regions are the most important (Table 4). Especially in Minas Gerais, Goiás, and São Paulo, the tropical climate is much characteristic with a hot and humid summer and a dry winter. These periods are characterized by different amount and quality of forages in pastures that affect the milk production directly, mainly in the systems that are simpler without or with low food supplementation during the dry period (Zocal and Carneiro, 2008).

Table 4. Milk production in Brazil, 2007*

Geographic regions	Volume (million of liters)	Participation in total (%)
Southeast	10.005	38
South	7.495	28
Mid-west	3.775	14
Northeast	3.460	13
North	1.737	7

*Estimating from Embrapa (CNPGL).
Source: IBGE – Pesquisa Pecuária Municipal (2008).

In Brazil, Minas Gerais state, located in Southeast region is responsible for almost 30% of whole milk production in Brazil. Other states such as Goiás, located in Mid-west region and Rio Grande do Sul in the South region has increased the milk production in the last years.

Brazil and Its Insertion in International Market of Milk and Dairy Products

After the 90th, significant changes occurred in the Brazilian economy related to milk supply chain and Brazil began to practice a model that inserted it in international economy (Gomes, 2002).

Since the end of the 90th, Brazilian economy has shown annual deficits of approximately US$ 450 million in commercial balance of dairy products. In spite of these results, in 2004, for the first time in Brazilian history, it was noted a surplus of almost 11.5 million of dollars (Table 5) (EMBRAPA, 2008).

The international market moves annually approximately 5% of the world milk production (Alvim *et al.,* 2002). According to Carvalho *et al.,* (2005), among the reasons that explain this low quantity marketed among the countries, it can be distressed: dairy products are considered one of the most protected products around the world with almost US$ 40 billion in subsidies per year according to OECD (Organization for Co-ooperation and Economic Development); milk is produced in several regions of the world, under dry and moisture climates, in high and low latitudes; its consumption has been increased in countries that have also increased their production. The authors emphasize that for each more 100 liters of milk produced in market in 2010, 71 liters would be from countries in development, most for local consumption.

Table 5. Brazilian commercial balance of dairy products

Economic activity	Year					
	1999	2000	2001	2002	2003	2004
Exportations	7.520	13.361	25.030	40.246	48.508	95.381
Importation	439.948	373.100	178.606	247.210	112.292	83.925
Balance	-432.428	-359.739	-153.576	-206.964	-63.784	11.456

Fonte: EMBRAPA (2008).

According to data studied by Carvalho *et al.* (2005), Brazilian milk has less solids content when compared with milk from other countries. While in Brazil, 8.2 liters of milk are necessary to produce 1 Kg of powdered milk, in New Zealand, 6.8 liters of milk are used to produce the same quantity of powdered milk. Although improvements in milk production can be noted in different regions of Brazil, it is also necessary improve milk quality, productivity, and sanitary conditions of dairy herds.

Predictions in World Milk Production until 2018

The Food and Agricultural Policy Research Institute (FAPRI) states that the world milk production will increase 18.2% in the next decade and the majority of this growth will be due to improvements related to productivity by cow. The world production projected to 2016 is 597.7 million of ton of milk (FAPRI, 2008).

It is estimated that 92 million of tons added to milk production from 2007 to 2016 will occur in America and 55% in Asia, mainly in China and India. Asia will be the greatest milk producer in 2016 with 195.8 million of tons, followed by Europe with 137.6 million of tons and by the United States of America, whose projected production is 110.45 millions of milk tons. According to FAPRI, India will produce 64% of Asia production in 2016. In North America, the United States will be the greatest milk producers and in the

South America, Brazil will produce 34.2 million of milk tons in 2016 and Argentina will be responsible for 14.4 million of tons (Brasil, 2008).

According to predictions from the Assessoria de Gestão Estratégica (AGE) at the Brazilian Ministry of Agriculture, Livestock and Food Supppy (MAPA), milk production in Brazil must grow at an annual tax of 1.92% in period from 2007/2008 to 2017/2018. The Brazilian milk production predicted to 2017/2018 is 33.1 million of tons. This amount will represent an added of 6.41 million of tons when compared to 2006/2007 period. The consumption will grow at an annual tax of 1.84% during this period, reaching 32.4 million of tons in 2017/2018 (Brasil, 2008).

In other prediction from the Brazilian Ministry of Agriculture Livestock and Food Supply, it can be noted that in 2019/2020 (Table 6), milk production can increase and reach 42.86 billion of liters (Brasil, 2010). If it occurs, Brazil probably will be one of the three greatest milk producers in the world.

Table 6. Milk production in Brazil from 2008 to 2020

Period	Production (billion of Liters)		
	Projection	Lower limit	Higher limit
2008/09	30.34		
2009/10	31.12	30,09	32.16
2010/11	31.80	29.97	33.63
2011/12	32.46	30.04	34.88
2012/13	33.12	30.23	36.042
2013/14	33.78	30.48	37.09
2014/15	34.45	30.78	38.11
2015/16	35.11	31.11	39.10
2016/17	35.77	31.47	40.07
2017/18	36.43	31.85	41.01
2018/19	37.09	32.24	41.94
2019/20	37.75	32.65	42.86

Elaborated by AGE/MAPA with data from LSPA/IBGE, USDA and EMBRAPA.
Note: Values between parentheses are considered at 95% of confidence interval.
Source: Brasil (2010).

2. Consumption

Milk is the first food offered to humans and their consumption remains important throughout life.

Table 7. Per capita consumption of milk in selected countries from 2000 to 2008

Country	kg / per capita / year								
	2000	2001	2002	2003	2004	2005	2006	2007	2008*
NORTH AMERICA									
Canada	93.1	92.1	90.4	87.2	86.9	86.4	93.9	93.9	92.6
United States	95.2	94.2	93.9	94.3	93.8	93.2	92.0	95.6	97.6
Mexico	39.2	40.2	39.8	42.0	41.4	42.1	40.9	42.1	42.1
SOUTH AMERICA									
Argentina	61.3	62.0	51.9	52.9	46.0	48.1	48.6	51.1	53.7
Brazil	72.3	69.7	68.3	68.1	69.2	70.8	72.7	77.0	83.2
EUROPE UNION**									
EUROPE UNION**	80.0	80.2	75.8	76.0	75.2	73.7	69.3	69.1	69.1
ORIENTAL EUROPE									
Romania	153.0	156.0	154.4	163.6	171.8	165.7	171.2	-	-
EX – URSS									
Russia	96.5	96.8	98.8	92.3	89.6	86.8	83.8	83.8	85.2
Ukrainian	63.3	66.0	68.7	72.4	108.0	91.9	105.9	109.0	109.5
ÁFRICA									
Egypt	18.2	21.5	21.1	21.8	21.5	21.2	20.8	-	-
ÁSIA									
China	3.0	3.5	4.4	5.9	7.9	9.9	10.4	11.2	12.0
South Korea	-	-	34.7	37.9	33.1	32.1	31.8	-	-
Índia	32.9	32.7	32.4	32.4	33.3	35.6	34.7	35.7	37.1
Japan	39.2	38.9	39.4	39.6	38.9	37.7	37.3	-	-
Taiwan	15.3	15.5	14.7	15.3	14.5	14.4	14.1	-	-.
OCEANIA									
Australia	103.9	99.2	100.6	100.4	101.4	103.7	103.6	105.3	108.5
New Zealand	90.6	91.9	90.8	91.1	90.1	89.2	87.0	87.0	87.0

Source: United States Department of Agriculture (2008), cited by EMBRAPA.
* Estimated.
** Europe Union – 27 countries.

Cow's milk and dairy products are widely consumed by human children and adults well after the age of weaning (Du et al., 2004).

Since mid-1990th the world milk consumption is growing in average 10 to 15 million t milk per year, based on the population growth and increasing income in many countries (IFCN, 2008)

Researches with humans have shown that high milk consumption is associated with a 10%–20% increase in circulating IGF-I levels among adults and a 20%–30% increase among children (Ma et al., 2001).

Milk is the most important source of calcium and vitamin D and therefore might be expected to decrease osteoporotic bone loss and fracture risk (Feskanich et. al., 2003).

Dairy consumption is a dietary factor that might affect type 2 diabetes. Several researches have suggested that dairy products may have favorable effects on body weight, the most important determinant of type 2 diabetes (Davies et al., 2000). Dairy intake may protect against type 2 diabetes by favorably affecting known risk factors or precursors of the disease (Choi et al., 2005).

Studies with dairy food consumption and stroke indicate that a higher intake of dairy foods reduces risk of stroke incidence. It is correlated with milk's minerals Ca, Mg and K (Massey, 2001).

The milk consumption per person varies widely. The Table 7 demonstrates milk per capita consumption from several countries of the world, since 2000 until 2008 (estimated). The consumption in developing countries has grown, such as in Brazil and Egypt. In developed countries, the consumption has stabilized, as Canada, United States, and Australia and decreased in other countries such as Europe Union countries and Russia.

The World Health Organization recommends the consumption, on average, of 600mL/Day or 219 liters/year, in fluid milk or dairy products. Milk consumption is directly related to per capita income of population. Since the GDP (Gross Domestic Product) has grown and improves income distribution in the country, increases the consumption of dairies by the population. In the next topic, details about the health benefits of milk.

3. Health Benefits

Milk can be considered as one of the most complete foods available, in nutritional terms. As well as being a great source of calcium, providing half of a child's (4-6 year olds) daily calcium requirement, a serving of milk contains important proteins, vitamin B2 and B12 and carbohydrate.

The health benefits of milk have been known since medieval times. Drinking milk has taken the advantage of the extensive nutritional value not only to the child, but also to the adult and the elderly. The benefits are the result of biologically active components that are present in native milk and also, due to their suitably modulated activities produced through the action of lactic bacteria (Santosa et al, 2006).

Among the macromolecules from milk, are carbohydrates (lactose), fats and protein and the micromolecules, such as biotin, iodine, Magnesium, Potassium, Pantothenic Acid, Riboflavin, Selenium, Thiamine and vitamins A, B, D and K (USDA National Nutritional Database for Standard Reference).

The lipids in milk are emulsified in globules covered with membranes. The proteins are in colloidal dispersions as micelles. The casein micelles happen as colloidal complexes of protein and salts, primarily calcium Lactose and the most of minerals are in solution. Specific milk proteins are involved in the early development of immune response, where others take part in the non-immunological defense, such as lactoferrin. All these components become milk a nutrient rich food (Keenan and Patton, 1995).

Many dairy products containing prebiotics or probiotics claiming to have a total range of health benefits are appearing on the market. Can be cited the microbiota of the probiotic competes with pathogens to colonize the intestine and stimulate the immune system. Bacteriocins, organic acids, and hydrogen peroxide produced by probiotics inhibit pathogens. Probiotics can be immunomodulators (Champagne et al., 2005, Coconnier et al., 1993).

3.1. Lactose

Lactose is a disaccharide sugar that is found exclusively in mammalian milk and is digested by the enzyme lactase in the mucosal brush border of the intestine. Reduced intestinal lactase results in malabsorption of lactose. The unabsorbed lactose is metabolized by colonic bacteria to produce gas and short chain fatty acids, causing the clinical syndrome of abdominal cramps, bloating, diarrhoea, and flatulence.

The enzyme β-galactosidase, more commonly known as lactase, is a responsible for the hydrolysis of lactose to the monosaccharides, glucose and galactose. These are absorbed by intestinal enterocytes into the bloodstream, glucose so is used as a source of energy and galactose becomes a component of glycolipids and glycoproteins.

Lactase deficiency may be classified as primary, secondary, congenital, and developmental. The classification is important as it relates to diagnosis, prognosis, and treatment (Bhatnagar and Aggarwal, 2007). Congenital lactase deficiency is related with the least lactase activity. It is a lifelong illness characterized by failure to thrive and infantile diarrhea from the first exposure to breast milk. Extremely rare, with only around 40 cases having been reported. It is a single autosomal recessive disorder, but very little is known about the molecular basis (Swallow, 2003). Just one treatment is complete avoidance of lactose from birth. Primary lactase deficiency is the most common cause of lactose intolerance. This type of lactase deficiency is genetically inherited and usually develops between the ages of two and 20. It´s develops when your lactase production decreases as a result of your diet being less reliant on milk and dairy products. This is more common after the two years old, when breastfeeding or bottle-feeding has stopped, though the symptoms may not be noticeable until many years later (McBean and Miller, 1998). Secondary hypolactasia can be the consequence of any condition that damages the small intestinal mucosa brush border or significantly increases the gastrointestinal transit time. Thus, secondary hypolactasia is transient and reversible (Labayen et al., 2001).

In patients with lactase nonpersistence, it´s common exclude milk and dairy products from the diet. However, this strategy may have serious nutritional disadvantages, chiefly for reduced intake of substances such as calcium, phosphorus and vitamins, and may be associated with decreased bone mineral density (Di Stefano et al., 2002). Several studies have been carried out to find alternative approaches, such as exogenous β-galactosidase, yogurt and probiotics for their bacterial lactase activity, pharmacological and non pharmacological strategies that can prolong contact time between enzyme and substrate delaying gastrointestinal transit time, and chronic lactose ingestion to enhance colonic adaptation (Montalto et al, 2006).

3.2. Calcium

Milk is a primary source of calcium and vitamin D. Milk and dairy products provide an excellent source of bioavailable calcium and avoidance leads to a lower calcium intake, which is associated with reduced bone mineral density and an increased risk of developing osteoporosis (Jackson and Savaiano, 2001).

At low-to-moderate calcium intakes, vitamin D is indispensable for calcium absorption. Vitamin D is essential for calcium uptake and bone development and remodeling. The primary source of vitamin D is conversion in the skin, via exposure to UVB radiation (Holick, 1996).

3.3. Vitamins

Among the water soluble vitamins, cow milk contain thiamin (vitamin B1), riboflavin (vitamin B2), niacin (vitamin B3), pantothenic acid (vitamin B5), vitamin B6 (pyridoxine), vitamin B12 (cobalamin), vitamin C, and folate. Milk is a source of thiamin, riboflavin and vitamin B12. In small amounts, milk contain niacin, pantothenic acid, vitamin B6, vitamin C, and folate and is not considered the main source of these vitamins in the diet. Also, milk contain the fat soluble vitamins A, D, E, and K. Milk contains small amounts of vitamins E and K and is not considered the main source of these vitamins in the diet (Fox and McSweeney, 1998).

Among the health benefits from milk vitamins, vitamin A is required for good vision, immunological system and for regular growth and development of body tissues. Vitamin D presents an important function in the absorption of calcium and phosphorus and is essential for healthy bones and teeth. Vitamin E has an important role in preventing damage to structures such as cell membranes. Substances which prevent damage in this way are called anti-oxidants and have been linked with reducing the risk of diseases such as cancer and other degenerative diseases. Vitamin K is essential for correct blood clotting, but there is little or no vitamin K naturally found in milk. Vitamin B12 is required for maintenance of healthy nerves and red blood cells, energy production and normal cell division. Thiamin (vitamin B1) is required for carbohydrate metabolism, neurological and cardiac function. Riboflavin (vitamin B2) is necessary for the release of energy from foods and healthy membranes and skin. Niacin is involved in energy metabolism. Folate is an important vitamin essential for cell division and correct development of tissues. Pyridoxine (Vitamin B6) is an essential vitamin implicated in protein metabolism and is required for the formation of red blood cells and for maintaining a healthy immune and nervous system. Vitamin C is required for the correct structure and maintenance of blood vessels, cartilage, muscle and bone (Philips et al., 2003; Zemel, 2005; Parodi, 2005 and Liu, 2006).

3.4. Minerals

In accordance to USDA National Nutritional Database for Standard Reference, milk contains calcium, copper, iron, magnesium, manganese, phosphorus, potassium, selenium, sodium and zinc. Milk is a good source of calcium, magnesium, phosphorus, potassium, selenium, and zinc. The others aren't considered the main source of these minerals in the diet.

Phosphorus is essential for healthy bones and teeth as well as cell membrane structure, tissue growth and regulation of pH levels in the body. Magnesium is essential for skeletal development, protein synthesis, muscle contraction and nerve function. Zinc is a constituent of many enzymes in the body and help to fight infections and growth development, too. Potassium is mainly present in the fluid of the cells in the body and is important for fluid balance, muscle contraction, nerve conduction as well as for the correct functioning of the heart (Flynn, 2003; Theobald, 2005; Zemel, 2005).

REFERENCES

Alvim, R. S.; Martins, M. C.; Muestefaga, P. S. Desempenho da cadeia produtiva do leite no Brasil – visão dos produtores. In: Duarte, V. (Ed.) *O agronegócio do leite e políticas públicas para o seu desenvolvimento sustentável*. 1. ed. Juiz de Fora: Embrapa Gado de Leite, p.195-204, 2002.

Bhatnagar, S. and Aggarwal, R. Lactose Intolerance. *British Medical Journal*. 2007; 334(7608): 1331–1332.

Brasil. Ministério da Agricultura, Pecuária e Abastecimento. Assessoria de Gestão Estratégica – AGE. *Projeções do Agronegócio Mundial e Brasil 2006/07 a 2017/18*. Disponível em: < http:// agricultura.gov.br >. Acesso em 15 de maio de 2008.

Carvalho, M. P.; Alvim, R. S.; Martins, M. C. Considerações sobre a inserção do Brasil no mercado mundial de lácteos. In: Zoccal, R. (Ed.) *A inserção do Brasil no mercado internacional de lácteos*. 1 ed. Juiz de Fora: Embrapa Gado de Leite, p.39-56, 2005.

Champagne, C.P.; Gardner, N.J.; Roy, D. Challenges in the addition of probiotic cultures to foods. Crit. Rev. Food Sci. Nutr., v.45, p.61-84, 2005.

Choi, H. K., Willett, W. C., Stampfer, M. J., Rimm, E., Hu, F. B. Dairy Consumption and Risk of Type 2 Diabetes Mellitus in Men. *Arch Intern Med* 2005; 165:997-1003.

Coconnier, M. H., Bernet, M. F., Kerneis, S. Chauvi-ere, G., Fourniat, J. Servin, A. L. Inhibition of adhesion of enteroinvasive pathogens to human intestinal Caco-2 cells by *Lactobacillus acidophilus* strain LB decreases bacterial invasion. *FEMS Microbiol Let* 110: 299-306, 1993.

Davies, K. M., Heaney, R. P., Recker, R. R., Lappe, J. M., et al. Calcium intake and body weight. *J Clin Endocrinol Metab*, 2000, 85:4635-4638.

Di Stefano, M., Veneto, G., Malservisi, S., Cecchetti, L., Minguzzi, L., Strocchi, A., Corazza, G. R. Lactose Malabsorption and Intolerance and peak bone mass. *Gastroenterology*, 2002; 122: 1793-1799.

Du, X., Zhu, K., Trube, A., Zhang, Q., Ma, G., Hu, X., Fraser, D. R. Greenfield. H.: School-milk intervention trial enhances growth and bone mineral accretion in Chinese girls aged 10–12 years in Beijing. *British Journal of Nutrition*. 2004. 91:1-11.

EMBRAPA. Empresa Brasileira de Pesquisa Agropecuária. Centro Nacional de Pesquisa de Gado de Leite. Brasil. In www.cnpgl.embrapa.br . 15/03/2011.

EMBRAPA. Empresa Brasileira de Pesquisa Agropecuária. Centro de inteligência do leite, 2008. Disponível em: < http://www.cileite.com.br/>. Acesso em 3 de agosto de 2009.

FAPRI. World Agricultural Outlook 2008. Center for Agricultural and Rural Development-Iowa State University, 2008. Available in: <http://www.fapri.iastate.edu/publications>.Accessed in March, 2008.

Feskanich, D.; Willett, W. C.; Colditz, G. A. Calcium, vitamin D, milk consumption, and hip fractures: a prospective study among postmenopausal women. 2003. Am J Clin Nutr, 77:504–11

Flynn, A. The role of dietary calcium in bone health. *Proceedings of the Nutrition Society*, 2003, 62; 851-858.

Food and Agricultural Organization. FAO Statistics Division 2009. Disponível em: <http://faostat.fao.org>. Acesso em 3 de agosto de 2009.

Food and Agricultural Organization of the United Nations. FAO Statistics Division 2010. Accessed in 04/08/2011. http://www.fao.org/economic/ess/ess-publications/ess-yearbook/ess-yearbook2010/yearbook2010-production/en/

Fox, P. F., and P. L. H. McSweeney. *Dairy Chemistry and Biochemistry*. 1998. Blackie Academic and Professional, an imprint of Chapman and Hall, London.

Gomes, S. T. Situação atual e tendências da competitividade de sistemas de produção. In: Duarte, V. (Ed.) *O agronegócio do leite e políticas públicas para o seu desenvolvimento sustentável*. 1. ed. Juiz de Fora: Embrapa Gado de Leite, p.67-81, 2002.

IBGE – Instituto Brasileiro de Geografia e Estatística. Pesquisa pecuária municipal 2006. Disponível em: <http://www.ibge.gov.br>. Acesso em 15 de maio de 2008.

Holick, M. F. Vitamin D and bone health. *J Nutr*. 1996; 126 :1159S –1164S.

IFCN (2008): Dairy Report, online publication. www.ifcnnetwork.org, 25.05.2009.

International Dairy Federation. IDF/FIL: Bulletin 423, 2007.

Jackson, K. A., Savaiano, D. A. Lactose maldigestion, calcium intake and osteoporosis in African, Asian and Hispanic-Americans. *J Am Coll Nutr,* 2001; 20 (Suppl. 2): 198S–207S.

Keenan TW, Patton S: *The structure of milk.* In Handbook of milk composition Edited by: Jensen RG. Academic Press, USA; 1995:5-50.

Labayen I., Forga L., Gonzalez A., Lenoir-Wijnkoop I., Nutr R., Martinez J. A. Relationship between lactose digestion, gastrointestinal transit time and symptoms in lactose malabsorbers after dairy consumption. *Aliment Pharmacol Ther*, 2001; 15: 543-549.

Liu, S.; Choi, H. K.; Ford. E.; Song, Y.; Klevak, A.; Buring, J. E.; Manson, J. E. A Prospective Study of Dairy Intake and the Risk of Type 2 Diabetes in Women. *Diabetes Care*, 2006, 29:1579-1584.

Ma, J., Giovannucci, E. Pollak, M., Chan. J. M., Gaziano, J. M., Willett, W. Stampfer, M. J. Milk intake. circulating levels of insulin-like growth factor-I. and risk of colorectal cancer in men. *J Natl Cancer Inst.* 2001. 93:1330-1336.

McBean, L. D., Miller, G. D. Allaying fears and fallacies about lactose intolerance. *J. Am. Diet. Assoc.* 1998; 98:671–6.

McKenna MJ. Differences in vitamin D status between countries in young adults and the elderly. *Am J Med* ,1992; 93:69–77.

Martins, P. C. Oportunidades e desafios para a cadeia produtiva do leite. In: Zoccal, R. (Ed.) *A inserção do Brasil no mercado internacional de lácteos*. 1 ed. Juiz de Fora: Embrapa Gado de Leite, p.11-30, 2005.

Massey, L. K. Dairy Food Consumption, Blood Pressure and Stroke. *Journal of Nutrition,* 2001; 131:1875-1878.

Montalto, M., Curigliano, V., Santoro, L., Vastola, M., Cammarota, G., Manna, R., Gasbarrini, A., Gasbarrini, G. *World J Gastroenterol*, 2006; 12(2):187-191.

Parodi, P. W. Dairy Product Consumption and the Risk of Breast Cancer. *Journal of the American College of Nutrition*, 2005, 556S-568S.

Philips, S.M.; Bandini, L.G.; Cyr, H.; Colclough-Douglas, S.; Naumova, E.; Must, A. Dairy food consumption and body weight and fatness studied longitudinally over the adolescent period. *Int J Obes Relat Metab Disord*, 2003; 27; 1106-1113.

Santosa, S., Farnworth, E., Jones, P. J.: Probiotics and their potential health claims. *Nutr Rev*, 2006, 64:265–274.

Swallow DM. Genetics of lactase persistence and lactose intolerance. *Annu Rev Genet*,2003, 37:197–219.

USDA. Access in Mar/10/2011. Consulting http://www.ars.usda.gov/main/site_main.htm?modecode=12-35-45-00.

Theobald, H. E. Dietary Calcium and health. *Nutrition Bulletin*, 2005, 30; 237-277.

World Health Organization. Obesity: Preventing and Managing a Global Epidemic. Report of a WHO consultation. Technical Report Series 894. WHO: Geneva, 2000.

Zemel, M. The role of dairy foods in weight management. *J Amer Col Nutr*. 2005, 24; 537S-5346S.

Zoccal, R.; Carneiro, A. V. Conjuntura atual do leite brasileiro. *Balde Branco*, São Paulo, p. 94 - 95, out., 2008.

In: Raw Milk
Editors: J. Momani, et al

ISBN: 978-1-61470-641-0

Chapter 5

Camel Milk as Therapeutic Alternative to Treat Diabetes; Comparison with Insulin

Amel Sboui[1,2*], Touhami Khorchani[1], Mongi Djegham[3] and Omrane Belhadj[2]

[1]Laboratoire d'Elevage et de la Faune Sauvage,
Institut des Régions Arides, Médenine Tunisie
[2]Laboratoire de Biochimie et Techno Biologie,
Faculté des Sciences de Tunis, Tunisie
[3]Laboratoire de Physiologie thérapeutique,
Ecole Nationale de Médecine Vétérinaire Sidi Thabet Tunisie

Abstract

This study was performed to evaluate the efficacy of camel milk on alloxan-induced diabetic dogs and to follow this effect in addition to Can-insulin®.

Four groups, composed of 4 diabetic dogs each, were used as follow: group 1 was getting camel milk, and group 2 treated simultaneous with camel milk and Can-insulin®, and group 3 received cow milk simultaneous with Can-insulin®. Group 4 contained clinically healthy

* Corresponding author: Amel SBOUI E-mail: amelsb6@yahoo.fr Address: Arid Land Institute, Livestock and Wildlife Laboratory Route Edjorf, Elfgè 4119, Medenine, Tunisia, Phone number: +216.75.633.005, Fax number: +216.75.633.006

animals and was used as control. Each dog received 500 ml of milk/day during five weeks.

After three weeks, group 1 showed a significant decline on blood glucose levels from 10.33 ± 0.55 to 6.22 ± 0.5 mmol/L, this improvement on glycemic control was accompanied to a significant decrease on total proteins concentrations (from 79.66 ± 2.11 to 63.63 ± 4.43 g/L). A significant decline of cholesterol levels (from 6.84 ±1.2 to 4.9 ± 0.5 mmol/L) was shown after only two weeks of treatment. The same result was illustrated on group 2 treated simultaneous with camel milk and Can-Insulin. In group 3 the effect of Can-insulin was well shown only on blood glucose levels during the treatment.

The investigation in this research was the beneficial effect of camel milk on diabetic dogs and its independence to the treatment with Can-insulin®.

Keywords: Camel milk, cow milk, alloxan, diabetes, dog.

INTRODUCTION

Diabetes mellitus is one of the gland endocrine diseases in Human and animal which involves the blood circulatory system. About 6.3% of world population lives with diabetes [1]. Diabetes mellitus is a chronic disorder of metabolism caused by an absolute or relative lack of insulin. It is characterized by hyperglycemia in the postprandial and or fasting state and in its severe form is accompanied by ketosis and protein wasting [2]. This metabolic disorder can be caused chemically using alloxan, streptozotocine; alloxan diabetes is caused by the selective pancreatic beta cell toxicity of this composite [3-4].

Several species were sensitive to alloxan toxicity such as rats, rabbit and dogs [5-6]. In modern medicine, no satisfactory effective therapy is available to cure diabetes mellitus, although it can be managed by insulin treatment. However, the pharmaceutical drugs used in diabetic therapy are either too expensive or have undesirable side-effects or contraindications [7]. Therefore the search for more effective and safer hypoglycaemic agents has continued to be an area of active research [8].

In arid regions and in the wilderness, camel milk is known for its usefulness to treat diabetes mellitus. For example, an Indian study reported a hypoglycemic effect of camel milk on diabetic rats [9].

In this context this research was conducted to study the effect of camel milk added or no with Caninsulin® on alloxan – induced diabetic dogs.

Alloxan-diabetic dog was used because it is a model of insulin deficiency and insulin resistance while simulating postprandial conditions in diabetic patients [1-10]. This animal model can be useful to study the diabetic deficiencies and helpful to veterinary and medical researches [1].

METHODS

Animals and diets: Twenty Clinically normal adult mixed-breed dogs were prepared for this experiment.

These dogs were housed individually in the Tunisian Veterinary Medicine School, Sidi Thabet. Animals were fed once daily with 350-400 g of commercial dry chow and 300 g of beef.

This food was given to all dogs daily in the morning after drinking milk. All animal were controlled when drinking milk to be sure that all the quantity given was consumed by the dogs. Water was available ad libitum for dogs throughout the duration of the experiment.

Induction of diabetes: After fasting for an overnight, dogs were injected by an intravenous administration of 65 mg of alloxan monohydrate (Sigma, Aldrich, Germany) / Kg of body weight [10].

Milk samples: Camel milk used during this study was obtained from a camel herd (camelus dromedarius) belonging the Arid Land Institute and cow milk was given from a Tunisian breed of cow housed in the Veterinary School of Medecine.

The two types of milk were used fresh without any treatment or dilution.

Before distribution of raw milk to the animal, the pH and acidity of the milk sample was checked to monitor the freshness of milk. The gross composition of the two types of milk was determined (fat, total proteins and total solids). Fat content was measured using the neusol method as indicated by Farah, 1996 and the total proteins concentration was determined by the Kjeldahl method using a nitrogen conversion factor of 6.36 [12]. Total solids were evaluated after drying at 105°C until a steady weight was achieved.

Experimental design: Five groups -composed by 4 dogs each- were used in this step; group 1: diabetic dogs treated with camel milk, group 2: diabetic dogs treated simultaneous with camel milk and Can-insulin®, and group 3: diabetic dogs treated with Can-insulin® in addition to cow milk, and group 4 consisted of diabetic dogs no treated and group 5 composed of healthy dogs used as control. Five hundred ml of milk was given to each dog daily during five weeks. Can-insulin® (Intervet, Nederland B.V) was injected as indicated in the notice: subcutaneously with (1IE / kg of body weight + 3IE) at drinking milk (500 mL for each dog daily).

The experiment was divided into two periods: the first consisted of four weeks in which, the animals were treated with milk and/or Can-insulin® and the second period lasting three weeks (weeks 5, and 6 and 7) to follow the variations of all analyzed parameters after stopping the milk / and or Can-insulin® treatment.

Blood samples and serum analysis: Blood samples were drawn 3 times per week from the radial vein with catheter system; these samples were divided in two tubes: one for blood glucose assay (enclose oxalate fluorure), the other for cholesterol, Triglycerides (TG) and total proteins measures.

Blood glucose concentration was measured by a glucose oxidase method (Biomaghreb®) using a spectrophotometer CECIL (CE 2041) at 505 nm. Cholesterol and triglycerides concentrations were determined by enzymatic methods (Biomaghreb®) using spectrophotometer at 505 nm. Total proteins levels were measured at 546 nm.

Urine analysis: A urine sample from each animal was analyzed- weekly during the trial- using Bayer reagent strips for urine analysis. The parameters followed in our study were: Glucosuria, and proteinuria and ketones.

Statistical analysis: The data were expressed as the mean ± SEM and represent the average values for the animals in the same group. Each analysis was repeated three times and the average was used to compare between treatments. These data were subjected to statistical analysis using SAS computer software (SAS institute, 1998) and the data were compared between and within the experimental groups.

This test combines ANOVA with comparison of differences between the means of the treatments at the significance level of $p < 0.05$.

RESULTS

Gross Chemical Composition of Milk

The pH and acidity of the camel milk provided to the animals were respectively 6.41 ± 0.18 and 16.87 ± 1.035°Dornic. These characteristics for the cow milk were as follows: 6.61 ± 0.24 for pH and 17.12 ± 0.64°Dornic.

The camel milk used during this study was rich in total protein (34.15 ± 3.11 g/L) and in total solids (119.43 ± 1.84 g/L) compared with bovine milk (30.5 ± 1.95 g/L for total proteins and 104.88 ± 4.39 g/L for total solid amounts). There was no significant difference in fat among the camel and cow milk used (34.5 ± 3.1 g/L in camel milk and 32.5 ± 2.12 g/L in bovine milk).

Effect of Milk and/ or Can-Insulin Treatment on Diabetic Dogs

Blood Glucose Levels

After drinking camel milk for four weeks, group 1 showed statistically significant decrease in blood glucose levels (from 10.33 ± 0.55 to 6.22 ± 0.5 mmol/L; p=0.028; figure 1), The hypoglycemic effect of camel milk on this group was significantly observed after 3 weeks of treatment illustrated by a non significant difference in comparison with the healthy group (figure 1). The same result was shown on dogs from group 2 (figure 1) and a non significant difference between groups 1 and 2 was revealed.

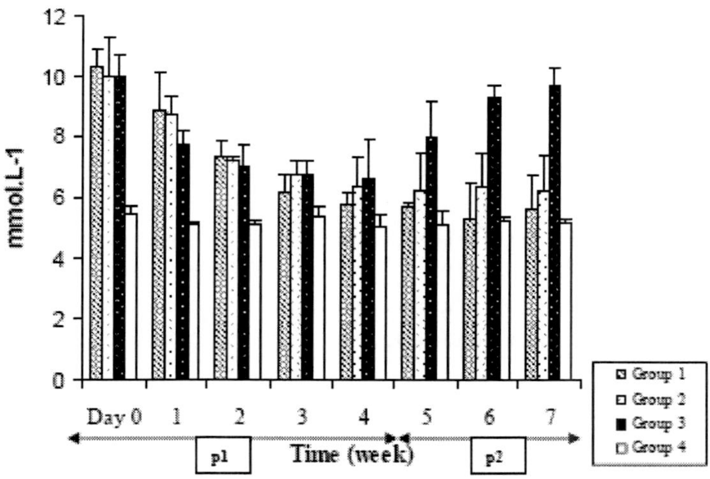

Legend of figure 1:

p1: period 1: Treatment with milk and / or Caninsulin®.

p2: period 2: After the end of the treatment (milk and/or Caninsulin®).

Group 1: Diabetic dogs treated with camel milk.

Group 2: Diabetic dogs treated with camel milk and Caninsulin®.

Group 3: Diabetic dogs treated with cow milk and Caninsulin®.

Group 4: Healthy group.

Figure 1. Weekly variations of blood glucose levels in groups 1, 2, 3 and 4 during the experiment.

During the trial, diabetic dogs from group 3 (treated with cow milk in addition to Can-insulin®) showed a significant decrease of blood glucose levels during the Can-Insulin treatment (from 10 ± 0.72 mmol/L to 6.66 ± 1.27 mmol/L, figure1). Once the Can-insulin® treatment was stopped (weeks 5, and 6 and 7), weekly variations of blood glucose levels showed a significant increase of this parameter (from 6.66 ± 1.27 to 9.72 ± 0.58 mmol/L).

During the period of testing, blood glucose levels in the healthy dogs (group 4) were within the normal range (3.33 – 6 mmol/L) (figure 1).

Total Proteins, Cholesterol and TG Variations

Only TG concentrations did not show any variations for all treatments during the experiment (table 1). In group 1, the improvement of glycemic balance after three weeks of camel milk treatment was accompanied to a significant decrease in total proteins concentrations (from 79.66 ± 2.11 g/L to 63.93 ± 2.61 g/L, table 2). A fast decline on cholesterol levels was shown after 2 weeks on this group (from 6.84 ± 1.2 mmol/L to 4.35 ± 0.61 mmol/L, table 1)

It was the same for the animal from group 2 (treated with camel milk and Caninsulin®); the weekly variations of these parameters demonstrated a non significant difference between groups 1 and 2.

Animals treated with cow milk in addition to Can-Insulin (group 3) showed a steady high cholesterol and total proteins concentrations during and after stopping of the treatment (about 7.23 ± 0.32 mmol/L for cholesterol and 81.49 ± 4.56 g/L for total proteins levels, table 1 and table 2).

The effectiveness of the treatment with camel milk supplemented or no with Caninsulin® (groups 1 and 2) was investigated on blood glucose, total proteins and cholesterol concentrations after the dogs stopped drinking milk (weeks 5, and 6 and 7); no significant differences were noted in outcomes analyzed (figure 1, table 1 and table 2) and all dogs showed a clinical healthy state by the end of the trial.

The non diabetic state was demonstrated in groups 1 and 2, firstly: by a normal range of fast blood glucose (5.66 ± 1.11 mmol/L), total proteins (64.63 ± 1.04 g/L), TG (1.02 ± 0.37 mmol/L) and cholesterol (4.11 ± 0.42 mmol/L) levels. Secondly: by the end of camel milk treatment all animal from groups 1 and 2 illustrated absence of glucose, proteins and ketones in urine sample which were well detected in urine sample after induction of diabetes.

This study was performed to evaluate the efficacy of camel milk (supplemented or no with Caninsulin®) in achieving glycemic control on Alloxan – induced diabetic dogs;

Alloxan injection causes a toxic effect on kidney and liver in addition to the pancreas as investigated by other study on alloxan induced- diabetes in [13-14]

Diabetes in dogs is generally associated, in addition to high blood glucose levels, to an increase of total proteins concentrations [4] which are illustrated in our study especially in dogs treated with cow milk (group 3) (82.83 ± 3.83 g/L).

Table 1. Weekly variations of cholesterol and TG levels in groups 1, 2, 3 and 4 during the trial

	Cholesterol (mmol/l)				TG (mmol/l)			
	Group 1	Group2	Group 3	Group4	Group 1	Group2	Group 3	Group 4
Day 0	6.84^{a}±- 1.2	6.94^{a} ± 0.5	6.7^{a} ± -0.5	4.17^{b} ±1.2	1.19^{a}± 0.27	1.03^{a}± 0.17	1.03^{a}± 0.27	0.95^{a}± 0.27
Week 1	6.9^{a} ± 0.25	6.58^{a} ± 0.85	6.9^{a}± 0.15	3.98^{b}± 0.25	1.21^{a} ±0.22	0.97^{b}± 0.19	0.82^{a}± 0.22	0.64^{a}± 0.22
Week 2	4.9^{b} ± 0.5	5.23^{b} ± 0.5	7.75^{a}± 0.07	4.7^{b}± 0.07	1.13^{a} ±0.15	0.9^{a,b}± .63	0.85^{a}± 0.15	0.66^{a}± 0.18
Week 3	4.92^{b} ±0.36	5.03^{b} ± 0.36	6.95^{a}± 0.1	4.82^{b}± 0.54	1.12^{a} ±0.15	0.94^{a}± 0.07	0.9^{a} ± 0.15	0.74^{a}± 0.15
Week 4	4.4^{b} ± 0.62	4.08^{b} ± 0.62	7.82^{a}± 0.46	4.21^{b}± 0.46	1.17^{a} ±0.08	0.97^{b}± 0.25	1.1^{a} ± 0.08	0.92^{a}± 0.83
Week 5	4.35^{b}± 0.61	4.33^{b} ± 0.61	7.13^{a}± 0.33	4.08^{b}± 0.33	0.99^{a} ± 0.3	0.94^{a}± 0.33	1.03^{a} ± 0.3	0.9^{a} ± 0.32
Week 6	4.27^{b} ± 0.5	4.44^{b} ± 0.64	7.34^{a}± 0.56	4.11^{b}± 0.62	1.05^{a} ±0.42	0.99^{a}± 0.52	1.01^{a}± 0.38	1^{a} ± 0.42
Week 7	4.11^{b}± 0.42	4.48^{b} ± 0.71	7.26^{a}± 0.36	4.6^{b}± 0.56	1.02^{a} ±0.37	1.13^{a}± 0.64	0.98^{a}± 0.62	0.97± 0.33

For each analyzed parameter: Means with the same letter in each line are not significantly different.

Group 1: diabetic dogs receiving camel milk.

Group 2: Diabetic dogs treated simultaneous with camel milk and Caninsulin®.

Group 3: Diabetic dogs treated simultaneous with cow milk and Caninsulin®.

Group 4: Healthy group receiving camel milk and used as control.

Week 1 to week 4: during the treatment.

Weeks 5 + 6 + 7: After stopping to drink milk and injection of Caninsulin®.

Table 2. Weekly variations of Total Proteins concentrations in groups 1, 2, 3 and 4 during the experiment

	Total proteins (g/l)			
	Group 1	Group 2	Group 3	Group 4
Day 0	79.66[a] ± 2.11	80.36[a] ± 0.9	79.18[a] ± 2.11	68.48[b] ± 2.01
Week 1	74.35[a] ± 7.25	71,93[a] ± 5.5	81.56[a] ± 7.25	68.8 [b] ± 3.25
Week 2	67.06[a,b] ± 9.91	67.35 [a,b] ± 7.13	82.45[a] ± 9.91	67.06[b] ± 2.27
Week 3	63.63[b] ± 4.43	66.05[b] ± 2.47	85.2[a] ± 4.43	65.75[b] ± 2.27
Week 4	64.58[b] ± 3.16	65.98[b] ± 1.77	74.76[a] ± 3.16	64.82[b] ± 2.11
Week 5	63.93[b] ± 2.61	63.14[b] ± 1.21	84.33[a] ± 2.61	65.45[b] ± 1.03
Week 6	66.57[b] ± 2	62.46[b] ± 2.35	82.09[a] ± 3.49	66 [b] ± 1.77
Week 7	64.63[b] ± 1.04	62.63[b] ± 3.14	82.36[a] ± 3.67	66.63 [b] ± 0.53

For each analyzed parameter: Means with the same letter in each line are not significantly different.

Group 1: diabetic dogs receiving camel milk.

Group 2: Diabetic dogs treated simultaneous with camel milk and Caninsulin®.

Group 3: Diabetic dogs treated simultaneous with cow milk and Caninsulin®.

Group 4: Healthy group receiving camel milk and used as control.

Week 1 to week 4: during the treatment.

Weeks 5 + 6 + 7: After stopping to drink milk and injection of Caninsulin®.

Some hypothesis [9-15] reported that the hypoglycemic effect of camel milk may be due to the high level of insulin in comparison with cow milk. But in this assay our results can not be due to this particularity because the effect of camel milk on glycemic control, proteins and lipids profile was observed also by the end of treatment (groups 1 and 2).

Hypoglycemic effect of Caninsulin was shown when it was injected with cow milk to the diabetic animals (figure 1). This effect was not illustrated when Caninsulin was injected to the diabetic animals treated simultaneous with camel milk. Caninsulin doesn't have any supplementary effect on the glycemic balance when added to camel milk (non significant difference compared with the effect of camel milk only). Camel milk may be able to eliminate the alloxan toxicity on pancreas or has a regenerative effect on beta cells and could be used as a curative treatment of diabetes in dogs.

High mineral content (Sodium, Potassium, Copper and Magnesium) as well as a high vitamin C intake [16] may act as antioxidant there by removing free radicals, which may provide an additional benefit to the animals treated with camel milk [17] It may be explained by the particularity and properties of camel milk in comparison with milk from other species, such as the absence of β-lactoglobulin, the high amount of polyunsaturated fatty acids (C18:1-C18:

3), and the high amount of vitamin B3 [18-19] and also some particularities of camel immunoglobulin, such as their small size and weight which offers enormous potential to camel milk. Also camel milk immunoglobulins, of relatively small size and weight, might offer interplay with host cell protein leading to an induction of regulatory cells and finally leading to a downward regulation of immune system and β-cell salvage [20-21].

From the results offered in our study, a therapeutic efficacy of camel milk on alloxan induced diabetes is showed. This may have important implication for the clinical management of diabetes mellitus in humans. But further studies are warranted to fractionate the active principle and to find out its exact mode of action.

References

[1] Valilou M, Sohrabi HI, Mohamednejad D, Soleimani RJ. Histopathological and ultrastructural lesions study of kidneys of alloxan induced diabetes mellitus in German Shephered dogs. *J. Animal and vet. Adv.* 2007; 6(8):1012 – 1016.

[2] Tyberg B, Anderson A, Hakan Borg LA. Species differences in susceptibility of transplanted and cultured pancreatic islets to the β- cell. *General Comparative Endocrinology* 2001; 122: 238-251.

[3] Rerup CC. Drugs producing diabetes through damage of the insulin secreting cells. *Pharmacol. Rev.* 1970; 2 : 485-518.

[4] Toulon F. Le diabète sucré du chien, maladie chronique. *Le Point Vétérinaire* 1986; 17 (94) : 681- 691.

[5] Sakudelski T. Mechanism of alloxan and streptozotocin action in beta cells of the rat pancreas. *Physiol. Res.* 2001; 50: 537-546.

[6] Stanely-Prince P, Kamalakkannan N, Menon VP. Antidiabetic and antihyperlipidaemic effect of alcoholic Syzigium cumini seeds in alloxan induced diabetic albino rats. *J. Ethnopharmacol* 2004; 91: 209–213.

[7] Lemhadri A, Zeggwagh NA, Maghrani M, Jouad H, Eddouks M. Anti-hyperglycaemic activity of the aqueous extract of Origanum vulgare growing wild in Tafilalet region. *J. Ethnopharmacol* 2004; 92: 251–256.

[8] Bell GI, Molecular defects in diabetes mellitus. *Diabetes* 1991; 40: 413-417.

[9] Agrawal RP, Sahani MS, Tuteja FC, Ghouri S.K, Sena DS, Gupta R, Kochar DK. Hypoglycemic Activity of Camel Milk in Chemically

Pancreatectomized Rats- An Experimental Study. *Int. J. Diab. Dev. Countries* 2005; 25 (3): 75-79.

[10] Matsuhisa M, Shi ZQ, Wan C. The effect of pioglitazone on hepatic glucose uptake measured with indirect and direct methods in alloxan – induced diabetic dogs. Diabetes 1997;46: 224-231.

[11] Farah Z. Camel milk: Properties and products, third ed. Swiss Centre for Development Cooperation in Technology and Management, St.Gallen, Switzerland 1996.

[12] Association française de normalisation (afnor). (1993) Contrôle de la qualité des produits alimentaires. Lait et produits laitiers, Afnor (Ed.), Paris, France.

[13] Kim JM, Chung JY, Lee SY, Choi EW, Kim M.K, Hwang C.Y, Youn HY. Hypoglycemic effects of vanadium on alloxan monohydrate induced - diabetic dogs. *J. Vet. Sci.* 2006; 7(4): 391–395.

[14] Pari L, Amarnath Satheesh M. Antidiabetic activity of Boerhaavia diffusa L: Effect on hepatic key enzymes in experimental diabetes. *J. Ethnopharmacol* 2004; 91: 109–113.

[15] Agrawal R., Swami SC, Beniwal R. Effect of camel milk on glycemic control, risk factors and diabetes quality of life in type 1 diabetes: A randomized prospective controlled study. *J. Camel Practice and Research* 2003; 10 (1): 45-50.

[16] Stahl HP, Sallmann R, Duehlmeir U, Wernery U. Selected vitamins and fatty acid patterns in dromedary milk and colostrums. *J. camel. Res. and Pract.* 2006; 13 (1): 53-57.

[17] Elsner M, Tiedge M, Lenzen S. Mechanism underlying resistance of human pancreatic beta cells against streptozotocin and alloxan. *Diabetologia* 2003; 46: 1713-1714.

[18] Farah Z. Composition and characteristics of camel milk. *J. Dairy. Res.* 1993; 60, 603-626.

[19] Zhang H, Yao J, Zaho D, Liu H, Guo M. Changes in chemical composition of Alxa Bactrian camel milk during lactation. *J. Dairy Sci.* 2005; 88: 3402- 3410.

[20] Hamers-Casterman C, Atarbouch, T, Muyldermans S, RobinsonnG, Bajyana Songa E, Hamers R. Naturally occurring antibodies devoid of light chains. *Nature* 1993; 363: 446-448.

[21] Rajendra P, Agrawal SS, Poornima S, Rajendra PG, Kochara DK, Mohan SS. Effect of camel milk on residual β-cell function in recent onset type 1 diabetes. *Diab. res. Clin. Prac.* 2007; 77(3): 494-495.

In: Raw Milk
Editors: J. Momani, et al

ISBN: 978-1-61470-641-0

Chapter 6

PROGRESS IN PASTEURIZATION PROCESSING OF RAW MILK: BACTERICIDAL EFFECT AND EXTENSION OF SHELF LIFE, IMPACTS ON THE PHYSICOCHEMICAL PROPERTIES, MILK COMPONENTS, FLAVOR AND PROCESSING CHARACTERISTICS

Ruijin Yang, Sha Zhang and Wei Zhao
State Key Laboratory of Food Science and Technology & School of Food Science and Technology, Jiangnan University, Wuxi, Jiangsu, China

1. CONVENTIONAL PASTEURIZATION OF MILK

Milk is a type of nutritionally complete food which contains protein, fat, lactose, vitamins, and minerals. The high nutritional content value of milk has become an excellent broth for a variety of microorganisms, which include many sorts of pathogens, such as (*Escherichia. coli*, *Listeria), (monocytogenes* and *Bacillus cereus*); (Fox and Cameron, 1982). The main purpose of pasteurization is to exterminate such pathogens in order to ensure the safety of milk and extend its shelf life. However, the pasteurization could also influence the physicochemical properties of milk, such as the changes of nutrient component which may reduce the digestibility and nutritional value of milk.

Meantime, the sensory quality of milk also decreased slightly due to the heat treatment.

1.1. Effects of Pasteurization on the Microorganisms in Raw Milk

Bacterial spoilage is one of the major factors in extending the shelf life of conventional pasteurized milk. Microbial growth and metabolism have subsequently shorten the shelf life of milk by producing undesirable changes in the aroma and taste attributes that influence consumer acceptability of the products (Frommand and Boor, 2004). The main bacteria of raw milk are (*Lactococcus lactis*, *Staphylococcus*, *E. coli& psychrotrophic bacteria)* (Cronjé, 2003). The pathogens of the raw milk were almost exterminated after pasteurization. However, some heat resistant bacteria still remained in the milk medium. The thermoduric bacteria are organisms capable of surviving the industrial pasteurization processing, and can be transferred into products causing quality defects, or creating health hazards. When pathogenic bacteria are relatively low in raw milk (less than 500 cfu/mL), some bacterial species did not only survive pasteurization, but grew in very large numbers during the food manufacturing processes. The heat resistant bacteria included (*lactic acid micro coli*, *streptococcus thermophilus)* and heat resistant micro (*aureus* & *endorspores* of *bacillus)*. The survival of these bacteria's caused huge problems for food manufacturers. Much time and efforts have been expended on the studying methods for controlling the large numbers of these bacteria in milk and milk products.

Many researches were focused on the effect of heat treatment condition to the resistance of bacteria. (Cronjé 2003) isolated and identified microbes in pasteurized and in double -pasteurized milk. The "milk isolates" included strains of (*Acinetobacter sp.*, *Candida lipolytica*, *Chryseobacterium meningosepticum*, *Pseudomonas putida)* and four isolates which is related to the *(Bacillus cereus)* group. The presence of these microorganisms in pasteurized milk can cause spoilage before the expiration date of the product .Their survival of the pasteurization is determined not only by their survival ability, but the pasteurization conditions. (Dumalisile et al. 2005) has investigated the different pasteurization conditions to the survival ability of these bacteria. It reported that different pasteurization methods (LTLT, HTST and pot pasteurization) placed different impacts on the sterilization of these milk bacteria. The research indicated that Bacterial strains of (*E. coli*, *A.*

baumannii, B. cereus, Chr. meningosepticum, P. putida,) yeast *(Can. Lipolytica)* and a reference strain (*B. coagulans*) were pasteurized by different pasteurization methods. Only the (*B. cereus)* strain could survive pasteurization in the LTLT and the HTST pasteurization treatments, whereas the other bacterial and yeast strains did not survive. By contrast, the same bacterial strains when treated with the 'pot' pasteurizer survived pasteurization, with the exception of the yeast. In short, different pasteurization methods showed different efficiency for the elimination of microorganisms. Furthermore, the microbiological quality of the raw milk before processing would place an impact on the final milk quality after pasteurization. Thus, there are different pasteurization standards for different dairy products, which depend on the bacteria quality of raw milk, fat content and the intended usage. (e.g.), the pasteurization standards for cream differs from the standards for fluid milk and the standards for pasteurizing cheese are designed to preserve the phosphatase, which aids in cutting. The HTST pasteurization standard was designed to achieve a 5-log reduction, killing 99.999% of the number of viable micro-organisms in milk. This is considered adequate for destroying almost all yeasts, mold and common spoilage bacteria and to also ensure adequate extermination of common pathogenic heat-resistant organisms (including *Mycobacterium tuberculosis*, which causes tuberculosis but not Coxiella burnetii, which causes Q-fever). HTST pasteurization processes must be designed appropriately so that the milk is heated evenly, and no part of the milk is subject to a shorter time or a lower temperature. (Champagne et al. 1994) has reviewed the growth and activity of psychrotrophs in milk. The psychrotrophic bacteria in milk do not cause the serious problems related to the spoiling of the milk (Čanigová, et al., 2002). It is well known that, Gram-negative bacteria, such as the (*Pseudomonas, Moraxella, Flavobacterium, Acinetobacter, & Alcaligenes)* predominate over Gram-positive bacteria in causing spoilage of pasteurized milks. These bacteria's are part of the micro flora of raw milk that resides in the dairy plant and contaminate the milk after it has been pasteurized because these Gram-negative bacteria are sensitive to heat and would be killed by normal pasteurization (Meer et al., 1991). In Canada, all milk produced at a processor and intended for consumption must be pasteurized, legally requiring it to be heated to at least 72 $^{o\ C}$ for at least 16 s and then cooling it to 4 $^{o\ C}$. This ensures the elimination of any harmful bacteria and the re growth of bacteria in the shelf life of milk. There are different temperatures for the pasteurization, but the shelf life of the milk will not be influenced by the process temperatures. (Gandy et al. 2008) have investigated the effect of pasteurization temperature on the shelf life of fluid milk. They found that varying pasteurization temperature had no effect on shelf-life. They also found that the milk could not be differentiated based on pasteurization temperature by a trained sensory descriptive panel or volatile compound composition toward the end of shelf-life. In addition, the shelf life of pasteurized milk was not only influenced by the pasteurization conditions but was affected by the packaging materials, due to post-pasteurization contamination which placed great impacts on the shelf life of milk. Meantime, (Petrus et al. 2010) have focused their research on the microbiological shelf life of pasteurized milk in bottle and pouch. They determined the Q_{10} and Z-value and presented that storage temperature has a greater effect on microbiological shelf life of pasteurized milk packaged in LDPE pouch compared to HDPE bottle. Thus, the HDPE bottles were preferred for its superior performance over the LDPE pouch with regard to microbial growth at storage temperatures ranging from 2 - 16 $^{o\,C}$ .In short, the factors limiting milk stability are well established: bacterial contamination, inadequate packaging system and improper temperature control. (Cromie 1991) reported the factors that influence the shelf life of pasteurized milk include the quality of the raw material, the binomial temperature/time pasteurization, resistant microorganisms to pasteurization (particularly psycrotrophics), the presence and activity of post pasteurization contaminants, the packaging system and storage temperature post pasteurization which had the greatest impacts on the stability of the product. (Griffiths & Phillips, 1990) reported that the one of the most critical factors lowering the durability of pasteurized milk products is the storage temperature of raw milk. (Burdova's) research indicated that storage temperature of 10 $^{o\,C}$ reduces the shelf life of pasteurized milk to one third in comparison with storage at 4.0 ^{o} C. The average shelf life of the full cream pasteurized milk reached 31 d at 4 ^{o} C; the average shelf life of skimmed pasteurized milk was 32.57 d.

Besides, (Douglas, 2000) have also published the result that the final microbial numbers were significantly influenced by the processing plant. (Fromm & Boor 2004) have also obtained the characterization method of pasteurized milk shelf life attributes. The Gram-positive organisms can be present in raw milk, but they also may enter milk products at various points during production and processing. They showed that the variability observed among plants suggests that plant-specific strategies will be needed to identify and reduce or eliminate sources of contamination. Development of these strategies might be achieved through systematic sampling of the dairy plant environment, including areas such as milk contact surfaces; equipment's, floors, and drains. Environmental sampling in place would facilitate to identify bacterial reservoirs, which must be targeted to reduce contamination at identified entry points and contribute to extended shelf life in fluid milk products. In other words, the control of the post pasteurization contaminants is as important as the pasteurization process on the microorganism quality of milk. In conclusion, the shelf life of pasteurized milk was affected by many aspects: the quality of the raw material, the binomial temperature/time pasteurization, resistant microorganisms to pasteurization, the presence and activity of post pasteurization contaminants, the packaging system and storage temperature post pasteurization. Currently, bacterial spoilage is still the most limiting factor in extending the shelf life of conventionally pasteurized high-temperature short-time (HTST) processed fluid milk products beyond 14 d (Boor 2001). However, the pasteurization still played an important role in the fluid milk processing, which provided adequate extermination of bacteria and offered full safety for human consumption.

1.2. Effects of Pasteurization on the Nutritional Quality of Milk

Milk is a rich source of vitamins, proteins and minerals which are important nutrients for the human being. Pasteurization, a kind of moderate heat treatment, definitely causes some damage to the nutrients in the milk although it may possibly guarantee safety for consumption of milk. For instance, HTST pasteurization, the most commonly used processing technique, is processing milk at greater than or equal to 72 $^{\circ}$C for greater than or equal to 15 s. The thermal instable nutrient will be damaged by the high temperature heating although with a short time. Milk a significant source of B vitamins, supplying thiamin, riboflavin, niacin, pantothenic acid, vitamin B_6, folate, and vitamin B_{12}. In the recent years, most of the researches were focused on the vitamins loss during the HTST heat treatments. The FDA contends that the major nutrients remain unchanged by pasteurization, and that thiamin, folate, B_{12} and riboflavin will experience losses from 0% - 10%. This reduction is described as "marginal" (Wong 1999; Miller et al., 2000). Figure 1 shows the comparison of vitamin loss due to the UHT and HTST processing. The thiamine, V_c, and B_{12} was damaged about 10% during the HTST processing while the other vitamins were not influenced by the HTST treatment. However, the UHT causes much severe reduction to the vitamins due to its high treatment temperature than the HTST, indicating that HTST is an effective method to reserve the vitamins. Riboflavin, another vitamin, is rich in the milk. It is heat stable and will not be affected by the hear treatment. However, the direct sunlight can cause the decrease of the riboflavin in the milk (Renner, 1986). Thus, the packaging material is a very important aspect to prevent the degradation of riboflavin. Folic acid another important vitamin which promoted the development of marrow cells (Bren et al. 2004) found that pasteurization induced less than 10% loss for the folic acid while the UHT damage more than 50% of the folic acid. They also showed that the addition of ascorbic acid can reduce the loss of folic acid during the heat treatment. Other researchers believed that the concentration of oxygen in the package will affect the loss the folic acid. Presently there is a folate binding protein which assists the intake of folic acid, some researches showed conflicting results about the protein damage during the pasteurization. The explanation of the conflicting results may be the HTST temperature is close to the denaturation temperature of this protein. (Anderson et al. 1994) studied the changes of vitamin B_{12}, folate and ascorbic acid during the storage. They found that there were no general or appreciable changes in vitamin B_{12} or folate content during storage. However, about 25–45% of the ascorbic acid was lost during storage. The levels of fat soluble vitamins, such as V_A, V_D and V_E in the milk, were slightly affected by the pasteurization processing due to the protection of the fat globules. In short, pasteurization temperature does not affect fat-soluble vitamins (A, D, and E), as well as the B-complex vitamins riboflavin, pantothenic acid, biotin, and niacin. The losses of vitamins are considered lower than those that take place during the normal handling and preparation of foodstuffs at home (Lund, 1982).

Figure 1. The vitamin loss due to the UHT and HTST processing. (Dairy Management Inc., 2003; Wong, 1999; Miller et al., 2000).

As it is well known, milk is an excellent resource of high biological value proteins due to the fact that milk can provide all of the essential amino acids that human being need. These essential amino acids can not be produced by the human body and must be ingested from the foods. The pasteurization promotes the Maillard reaction of the milk. The Maillard reaction can lead the degradation of the milk proteins and amino acids, thus reducing the protein quality of the milk. However, when compared with the UHT processing, the HTST processing causes much less reduction of the protein quality. (AlKanhal. 2001) pointed that this reduction in nutritional quality might be significant for children who are solely dependent on this type of milk in their diet. To some extent, heat treatment may denature milk proteins. This effect is not considered a disadvantage from the nutritional point of view because it only changes the specific arrangement of the casein. Since there are no breakdown of peptide linkages casein can be considered a thermal-resistant protein. Although α-lacto albumin is relatively heat stable, other whey proteins can be denatured by heating. These denatured proteins becoming more digestible than their naturally form in the milk because the protein structure is loosened and digestive enzymes can break them down easier (Renner, 1986). Milk contains a lot of antibacterial proteins (e.g.) lactoferrin, which binds free iron effectively limiting its availability to pathogens for growth, which is not affected by standard pasteurization techniques. Although ultra-pasteurization (UHT) does reduce its ability to bind free iron, bacteriocins and lysozyme, are not affected by pasteurization. Another milk protein, lactoperoxidase, contributes to the antibacterial properties of milk by catalyzing the production of hydrogen peroxide. Lactoperoxidase retains 70% of its activity and is heat stable even after pasteurization.

In addition, the pasteurization does not impair the nutritional quality of milk fat, calcium, and phosphorus (Beddows & Blake, 1982). The mineral of milk is rich and changed slightly during the pasteurization processing. Despite the fact that pasteurization may slightly reduce the amount of free calcium in the milk, both the total amount of calcium and the bioavailability of the calcium in milk remain unchanged after pasteurization. The iodine content of milk varies greatly, depending on the cow's condition, in some instances a 20% reduction in iodine has been reported with pasteurization. Furthermore, milk contains a large amount of lactose, about 4.8% to the total mass. Lactose in milk is stable during standard pasteurization; thus, the concentration of lactose in milk is not be significantly affected by pasteurization because the milk contains little alkali. Small amounts of lactose are transferred to the lactulose, a functional disaccharide which is useful to prevent constipation.

Nevertheless, pasteurization will destroy the lactase-producing bacteria that may be present in raw milk, which contribute to greater lactose tolerance. In addition, milk contains a lot of other nutritional components, such as oligosaccharides, lactoferrin, lysozyme, and lactoperoxidase, which are either unaffected or minimally affected by pasteurization. Oligosaccharides, as the bifidus factor binds to pathogens to prevent their adherence to target mucosal receptors, are heat stable. In conclusion, pasteurization does not significantly alter the nutritional value of milk.

2. Non-Thermal Technology Pasteurization of Milk

Thermal treatment is the most popular preservation technology for the elimination of microbial contamination of milk which guarantees the safety of milk, but applying heat to milk often causes cooked flavor and undesirable changes on its nutritional and physicochemical properties. Therefore, innovative and emerging non-thermal technologies, including High Pressure Processing (HPP), Pulsed Electric Fields (PEF), Ultrasonic Processing (UP), Microwave Processing (MP), Ultra Filtration (UF), which are able to inactivate microorganisms without undesirable heat, stand in the interest of scientists and food industry as an attractive preservation process for milk.

2.1. High Hydrostatic Pressure (HHP) Processing of Milk

The application of HHP in food preservation has received particular attention as an alternative to thermal processing. Milk was the first food product to be treated with HHP in nineteenth century (Hite, 1899). At present, it is commercially applied for a range of products, such as fruit juice, oysters, ready meals and meat product. HHP processing technology generally involves placing the product, with or without packaging, in a vessel, and applying the pressure through a piston or a pump for a desired time after the closure of the vessel. The achievable pressures generally range from 100-1000 MPa (Huppertz et al., 2006).

2.1.1 The Effect of HHP on Microorganisms of Milk

One of the principal advantages of the HHP process is that it can destroy microorganisms by high hydrostatic pressure without heat, expanding shelf life of food and improving food safety. The loss of viability of microorganisms through HHP is probably the result of a combination of injuries in the cell. HHP does not alter the low-energy, covalent bonds, the primary structure of molecules such as proteins or fatty acids remains intact, but HHP could alter the secondary, tertiary or quaternary structure of large molecules and complex organized structures such as membranes, so there is no single damage in a cellular structure, and the death of the cells is due to a multiplicity of damage accumulated in different parts of the cell (Rendueles et al., 2011).The effectiveness of the HHP treatment depends primarily on the pressure applied and on the holding time. The resistance of microorganisms is highly variable, depending on the processing conditions (pressure, time, temperature and cycles), food constituents, and physiological state of the microorganism (Smelt, 1998). Gram-positive bacteria are generally more resistant to HHP compared to Gram-negative. (e.g.),Gram- negative microorganisms need an application of 300-400 MPa at 25 $^{o\,C}$ for 10 min to achieve inactivation while Gram- positive microorganisms can be inactivated with 500-600 MPa with the same time and temperature (Chawla et al., 2011). The difference in resistance is due to the different chemical composition and structural properties of the cell membrane in Gram-positive and Gram-negative microorganisms. In addition, Bacteria cells at their exponential growing stage are more sensitive to HHP than in the stationary phase. The bacterial spores are the most resistant, and they can survive at pressure of 1000 MPa (Cheftel, 1992). However, spore can be inactivated by HHP along with mild heat treatment. The exact mechanism of spore inactivation in not well known, but it has been proposed that spores are first activated as a result of particular pressure/temperature conditions, losing their inherent resistance to pressure and heat, and subsequently get killed by treatment (Rendueles et al., 2011).

HHP is effective in destroying pathogenic and spoilage microorganisms. It has been reported that HHP treatment can inactivate 3 major food pathogens present in milk (*Listeria monocytogenes*, *Escherichia coli* and *Salmonella enteritidis*). Many researches have also proved that HHP treatment at 400 MPa for 15 min or 500 MPa for 3 min at ambient temperature can achieve a microbiological reduction similar to that of pasteurized milk (Vazquez-Landaverde et al., 2006). In order to achieve a shelf life of 10 days at a storage temperature of 10 $^{o\,C}$, a pressure treatment of 400 MPa for 15 min or 600 MPa for 3 min at 20 $^{o\,C}$ is needed (Rademacher et al., 1997). Furthermore, combination of HHP with heat treatment is a vital problem in extending the shelf life of milk. For example, a moderate temperature (55 °C) along with HHP (586 MPa for 5 min) can significantly extend the shelf life of milk beyond 45 days (Rademacher et al., 1997).

2.1.2. Effect of HHP on Milk Quality

Effect of HHP treatment on mineral balance in milk has been studied. The results indicate that HHP treatment results in two main features of mineral balance of milk, the level of ionized minerals, particularly calcium, and the distribution of minerals, primarily calcium and phosphate, between the micelles and the serum phase of milk (Huppertz et al., 2006). As a result of HHP treatment, the concentration of ionic calcium in milk increased, and the level of calcium and phosphate in the serum phase of milk also increased (Lopez-Fandino et al., 1998; Zobrist et al., 2005). The pH of milk slightly increased due to the increase of concentration of phosphate occurred in the milk serum (Schrader et al., 1998). During storage period, such increases are either reversible or irreversible, depending on the storage temperature. If milk was stored at 20 °C, the increases are rapidly reversible, while they are virtually irreversible on subsequent storage at 5 °C (Zobrist et al., 2005).

The effects of HHP treatment on milk proteins have become an area of considerable research interest in recent years, mainly including HHP-induced changes in casein micelles and whey proteins. When subjected to HHP treatment, the size, number, hydration, composition and light-scattering properties of casein micelles differ considerably from that in untreated milk (Huppertz et al., 2006). With the application of HHP treatment at 100-200 MPa at room temperature, the average size of casein micelles is comparable to those in untreated milk, but the micelles in milk treated at 250 MPa for more than 15 min are considerably larger than in untreated milk, while if the treatment pressure increase to > 300 MPa, the micelles are about 50% smaller than that in untreated milk (Huppertz et al., 2006). The HHP-induced increases in micelle size are because spherical particles change to form chains or clusters of sub-micelle (Huppertz et al., 2006). HHP treatment also influenced the number of casein micelle in milk considerably. The amount of sediment able protein at 100,000 × g in HHP-treated milk was less than that in untreated milk (Huppertz et al., 2004). The HHP-induced reduction in the level of sediment able casein is in agreement with HHP-induced increases in the level of caseins in the serum phase of milk. The hydration of casein micelles increases considerably by HHP treatment. There are two reasons to explain, one is that HHP-induced disruption of casein micelles into small particles, and the other one is that the association of denatured β-lg with casein micelles increases the net-negative charge on micelles (Gaucheron et al., 1997; Huppertz et al., 2004). The extent of light-scattering by β-casein micelles decreased with increasing pressure up to 150 MPa, but the extent of light-scattering progressively increased at a higher pressure (150-300 MPa) (Payens et al.,1969). These observations suggest that the hydrophobic bonds between casein molecules, in the main mechanism of micellisation in β-casein, are disrupted at pressure less than 150 MPa, but enhanced at higher pressures (Ohmiya et al., 1989).

HHP-induced denaturation of whey protein, primarily α-la and β-lg, is observed at pressures > 400 or >100 MPa, respectively (Huppertz et al., 2006a). The higher barostability of the α-la than β-lg might due to the absence of a free sulfhydryl group and higher number of intramolecular disulphide bonds in α-la. The extent of HHP-induced denaturation increases with increasing treatment time, treatment temperature, milk pH and the level of micellar calcium phosphate in the milk (Huppertz et al., 2006a). Furthermore, some denatured α-la and β-lg associated with milk fat globule membrane (MFGM) proteins via disulfide bonds during HHP treatment. The amount of β-lg associated with the MFGM increased with an increase in pressure and treatment time (Considine et al., 2007). In addition, the denaturation of whey proteins leads to interaction between denatured whey protein and casein, which results in modifying the technological parameters of milk to make cheese, improving the rennet coagulation properties and yield of cheese (Lopez-Fandino et al., 1998).

As for the effect of HHP on volatile profile of milk, HHP processing at low temperature causes minimum change of the volatile composition of milk. However, it has been found that pressure, temperature, and time, as well as their interactions, all had significant effects on volatile generation in milk. Pressure and time influences were significant at 60 ^{oC}, while their effects were almost negligible at 25 ^{oC} (Vazquez-Landaverde et al., 2006).

2.2. Pulsed Electric Field (PEF) Processing of Milk

Among non-thermal treatments, PEF has received special attention due to its feasible and energy efficient application in continuous-flow processing. PEF processing is conducted by introducing the food in a chamber which contain two electrodes to inactive the microorganisms by short high power electric pulses. Typical PEF system for the treatment of fluid foods consist of a pump, a PEF generation unit, which is composed of a high voltage generator and a pulse generator, a treatment chamber, a cooling device and a set of monitoring devices.

2.2.1. The Effect of PEF on Microorganisms of Milk

The PEF process is based on the fact that food usually contains ions, these will cause a current to flow through the food product which causes microbial inactivation by dielectrical breakdown and electroporation of cell membrane. When an external electric field is applied to a microbial cell, a Transmembrane potential is induced across the cell membrane. Then small metastable hydrophilic pores were created after the transmembrane potential has been built up. During the electric field treatment, the number of pores and their sizes changed, intracellular compounds leaked, and extracellular substances enter in the cell until the cell membrane loss its stability and functionality which lead to the death of microbial cell (Qin et al., 1996; Saulis, 2010). There are several theories to explain how pores are formed on the cell membrane but it is still unclear whether it occurs in the protein or lipid matrices (Barbosa-Gánovas et al., 1999), but the fact is that electric fields induce structural changes in the microbial cell membrane (Bendicho et al., 2002).

The level of microbial inactivation achieved with PEF treatment mainly depends on the process variables, such as electric field strength, pulsed width and frequency that applied during the process. In generally, the microbial inactivation markedly enhanced with the increased electric field strength and treatment time. It has also been reported that the enhancement of PEF effect led by certain combinations of the process variables. For example, the combined effect of the electric field strength and pulse width caused a greater reduction in the population of (*S. aureus*) in milk than the lethality achieved for each level of the variables when they were studied separately (Smith et al., 2002). Moreover, Microbial inactivation has also been related to the treatment temperature. It has been reported that an increase in treatment temperature leads to higher effectiveness in the inactivation of microorganisms. Heating skim milk from 13 to 33 $^{o\ C}$ accelerated the inactivation of (*Pseudomonas fluorescens* and *Listeria innocua*) as electric field strength, treatment time or energy input increased (Fernández-Molina et al., 2005).

The complex composition of milk has some influences on the efficiency of PEF treatment; it has been observed that the effectiveness of PEF treatment decreases in the raw milk in comparison with its action in dilute solutions and fruit juices (Otunola et al., 2008). Perhaps it's the complex composition of milk and high content of protein and fat may act as a shield to protect microorganisms from the lethal effect of PEF. In addition, the conductivity of milk is higher due to its charged compounds including mineral salts and bicarbonates (Lindgren et al., 2002), which results in shortening the pulse width which affected the degree of microbial survivability.

In raw milk, (*Escherichia coli*, *Staphylococcus aureus*, *Listeria monocytogens*) could be inactivated for 4 log cycles after a certain intensity of PEF treatment. However, difference type of microorganisms showed different resistance under PEF treatment. The reduction of microbial counts varied from 1 to more than 5 log cycles under the same strength of PEF treatment. It has been reported that (*Staphylococcus aureus)* and coagulase negative (*Staphylococcus sp*). could be inactivated 4 and 2 log cycles, respectively, while no reduction of other microorganisms such as (*Corynebacterium*)or(*Xanthomonas maltophilia)* was observed under the same PEF treatment (Raso et al., 1999). Differences in the degree of reduction in these microorganisms can be attributed to the differences in the size of the cells and the susceptibility of Gram-negative cells to PEF (Damar et al., 2002).

The shelf life of PEF-processed milk depends on the initial concentration of the PEF-resistant microorganisms, as well as on their ability to grow at refrigeration temperature. The PEF-processed milk was found to have a microbial shelf life of 2 weeks (Bendicho et al., 2002). However, the shelf life of milk could be extended if milk was treated by the combination of PEF with other methods, Such as moderate heating, nisin, and acetic or propionic acid. Particularly the combination of PEF with a mild thermal treatment has received much attention. (Fernández-Molina, Barbosa-Cánovas, & Swanson 2005) increased the shelf life of PEF-treated milk up to 30 days (stored by refrigeration) by applying a mild thermal treatment before the PEF process, which was equivalent to doubling the shelf life associated with any individually applied treatment.

2.2.2. Effect of PEF on Milk Quality

As one of the innovative non-thermal technologies, PEF has been shown mainly to preserve the nutritional components of food and minimally alter its sensory properties. No significant difference of physicochemical properties of milk, such as its viscosity, density, electrical conductivity, pH, protein and total solids content, was observed after raw milk was treated by PEF at a temperature below 52 $^{o\,C}$ (Martín et al., 1997). Furthermore, the concentrations of different fractions of whey proteins in milk were mildly reduced after PEF treatment without exceeding the temperature of 40 $^{o\,C}$, but still higher than that which was treated by traditional heat pasteurization (75 $^{o\,C}$, 15s) (Michalac et al., 2003). PEF affected milk coagulation properties, also PEF-treated milk showed better rennetability compared to thermally pasteurized milk, which indicate that PEF could be a potential substitute for pasteurization method for cheese making (Floury et al., 2006).

The changes of total concentration of fatty acids of milk were negligible processing by PEF; PEF processing could induce small globules to clump together, causing an apparent increment in the population of larger milk-fat globules (Garcia-Amezquita et al., 2009). There are also some studies showed that PEF treatment could induce hydrogen radical formation in treated samples, which in turn accelerate fat oxidation (Zhang et al., 2011). As a result, some volatile compounds of PEF-treated milk, mainly products of lipid oxidation were higher than that of untreated samples.

As for the effect of PEF on the vitamins in milk, no changes in thiamine, riboflavin, cholecalciferol and tocopherol contents were reported, whereas the ascorbic acid content of milk was reduced after PEF treatment following a first-order kinetic model (Bendicho et al., 2002). With regard to vitamin contents under storage at 4 ^{oC}, the stability of vitamins was similar irrespective of the treatment and technology applied except riboflavin, whose concentrations remained higher in PEF-treated samples than thermal treated milk after 15 and 60 days of storage at 4 ^{oC} (Sobrino-López et al., 2010).

It has been proven that thermal treatment alters sensory properties of milk, but PEF seems to keep nutritional content and sensory properties. PEF has been used to apply to retain the quality of milk destined for dairy products, such as cheeses, yogurt and milk beverage (Sobrino-López et al., 2010). Although the contents of some sensitive volatile compounds of PEF-treated milk differed from untreated samples, PEF processing can achieve a similar microbial inactivation than thermal processing with a better milk fresh aroma.

2.3. Ultrasonic Processing (UP) of Milk

The application of high intensity ultrasound processing (UP) in food industry is one of the merging alternate food processing technologies. Ultrasonication has been successfully used in the dairy industry for equipment cleaning and homogenization. Although the use of ultrasound to inactivate microbes was studied in the late 1920's (Harvey et al., 1929), its limitation in lethal effect on spoilage microbes prevented it from being used as a sterilization method. Thanks to the improvements in ultrasound generation technology, microbial inactivation by ultrasound has been again stimulated interest over the last decade. The advantages of application of UP in milk includes: homogenization of milk fat, remove of gas, minimal flavor losses, and substantial energy efficient.

2.3.1. The Effect of UP on Microorganisms of Milk

Ultrasound is defined as a sound wave with a frequency of above 20 kHz, which is above the frequency of human hearing. High intensity low frequency ultrasound, which is recommended for microbial inactivation, refers to ultrasound at frequencies of from 20 - 100 kHz (Mason et al., 2002). In general, the effect of ultrasound on microbial inactivation is attributed to the process is known as cavitation, which involves generation, growth, and collapse of bubbles (Gera et al., 2011). During ultrasonication, longitudinal sound waves are generated in the liquid medium, which in turn create regions of alternating compressions and rarefactions (Sala et al., 1995). The continuous pressure changes between the two regions lead to cavitation. Cavitation bubbles are formed in the rarefaction region and grow in size in the compression region until a critical size is reached, after which they are unable to sustain themselves and finally collapse violently by implosion. This collapse results in radiation of shock waves, which create micro-regions of very high temperature and pressure leading to microbial inactivation (Piyasena et al., 2003). However, the formation of free radicals and other reactive species during bubbles collapse, such as various species of oxygen and hydrogen peroxide, are commonly thought to be in the inactivation of microorganisms (Riesz et al., 1992). Therefore, the exact reason for the lethality of ultrasound has not been completely understood yet.

When ultrasound in applied in the food industry as a pasteurizing or sterilizing technology, there are a few critical processing factors that affect the efficiency of microbial elimination, including the amplitude of the ultrasonic waves, contact time with microorganisms, treatment volume, treatment temperature, the type and number of microbes to be treated and the composition of food (Hoover, 2000). It is generally assumed that the larger the microbial cells are, the more sensitive to the effects of ultrasound they will be. It has been reported that rods show more resistant when compared to coccoids, and aerobic microbes are more resistant than anaerobes. Gram-negative microbes have been found to be more sensitive to ultrasonication than Gram-positives. Spores are the most resistant ones to ultrasonication, and even questioned the ability of ultrasound to inactivate spores. In addition, the age of the cells is another important factor influencing sensitivity. For instance, young (4h) (*Saccharomyces cerevisiae)* cells were more sensitive than older ones (24h) (Kinsloe et al., 1954).

Ultrasound was found to eliminate spoilage and potential pathogens in milk to zero, even when initial inoculums loads of 5 times higher than permitted were present before UP treatment. It has been reported that viable cell counts of (*E. coli)* and (*Pseudomonas fluorescens)* in milk were reduced by 100% after 10.0 min and 6.0 min of ultrasonication, respectively, while (*Listeria monocytogenes)* in milk was reduced by 99% after 10.0 min (Cameron et al., 2009). Ultrasonication results in over 5 log reduction in total viable counts up to 6 days of storage (Chouliara et al., 2010).

Furthermore, high intensity ultrasound in conjunction with mild heating thermosonication and pressure manothermosonication for the inactivation of microbes has been received considerable interest. It has shown that the inactivation of (*Listeria innocua)* and (*mesophilic)* bacteria in raw milk is more efficient when thermosonication is used in place of purely thermal pasteurization (Bermúdez-Aguirre et al., 2009). Similarly, (Garcia 1898) found that the simultaneous use of heat (70 – 95 $^{o\,C}$) and ultrasound (20 kHz, 150 W) was more effective in the inactivation of (*Bacillus subtilis)* compared to individual treatment by heat or ultrasound alone. Thermosonication process is also effective for spores, which can reduce 70% -99.9% of the spores (Ashokkumar et al., 2010).

2.3.2. Effect of UP Processing on Milk Quality

Ultrasonication did not lead to decrease in protein or casein content of raw milk. However, it has been reported that ultrasonication disrupted casein micelles to generate free casein in the solution, but the reactive sulfhydryl content of the milk was not affected (Taylor et al., 1980). In addition, ultrasonication increased the water solubility of the whey proteins by about 5-6%. It has been suggested that ultrasonic treatment changed the conformation of the proteins leading to the exposure of hydrophilic moieties to water (Ashokkumar et al., 2010).

With regard to the effect of ultrasonication on milk fat, the studies showed that ultrasonication lead to an increase in the fat concentration, which can be explained by the larger surface area of the fat globules after ultrasonication (Cameron et al., 2009). Moreover, ultrasonication strongly induced free radical formation leading to enhanced lipid oxidation in milk, but malondialdehyde content of ultrasonic treated milk remained lower than the proposed limit, which constitutes a food product sensorial unacceptable due to lipid oxidation (Chouliara et al., 2010).

Although high intensity ultrasound has the potential to simultaneously homogenize milk and reduce its microbial load, the treatment may give rise to off-odors under certain conditions. According to sensory evaluation, researches described the off-odors after ultrasonication as "rubbery". And according to GC-MS analysis, volatiles generated by UP treatment in milk were predominantly hydrocarbons and believed to be of pyrolytic origin, possibly generated by high localized temperature associated with cavitation (Riener et al., 2009).

2.4. Microwave Processing (MP) of Milk

Microwave energy has been used since the early 1960s for several food processing operations such as cooking, baking and drying (Young et al., 1990). Since the first reported use of microwave system for pasteurizing milk in 1969 (Hamid et al., 1969), continuous microwave treatment has been proved to be an effective system for pasteurization of milk with several advantages including the speed of operation, energy savings, faster start-up and shut-down times.

2.4.1. Effect of MP on Microorganisms of Milk

The principle of heating with microwaves is very different from that of conventional heating by convection or conduction. Microwave are generated by a magnetron and then absorbed by the food present, and then the dipole molecules in food align with the microwave fields which cause friction among the molecules resulting in heating of food (Knutson et al., 1987). There is some controversy as to the exact microbial inactivation mechanism of microwaves. Some argued that microwave itself had a lethal effect, with no significant rise in temperature, on the microbes (Flemming, 1944), while others stated that microbial reduction was rather brought about by thermal effects and not the microwaves as much (Brown & Morrison, 1954; Lechowich et al., 1969; Vela & Wu, 1979). It is commonly accepted that heat, and not microwave radiation alone, kills the microorganisms.

When compared raw milk heated for 30 min in a continuous flow microwave heating system to raw milk heated for 30 min in a water bath at 63 $^{o\,C}$, both treatments achieved negative phosphatase tests, and no coli forms could be detected. A six log reduction was observed for plate counts (Merin et al., 1984). It has been reported that microwave heating could extend the shelf-life of pasteurized milk. Microwave heating of eight day old milk to 60 ^{o} C reduced the psychrotrophic microbial count (1.8×10^6 CFU.mL^{-1}) to zero, thus extending the shelf-life of milk (Chiu et al., 1984). In addition, raw milk treated by continuous flow microwave heating at 80 or 92 ^{o}C for 15 s could achieve a shelf-life of up to 15 days at 4 ^{o}C (Valero et al., 2000).

2.4.2. Effect of MP on Milk Quality

The effect of MP on milk vitamins has not found to be acceptable to researchers.

It has been claimed that destruction of vitamins in microwave heating treated milk was less than that in conventional processed milk. For intense, (Sieber et al. 1996) reported no loss of vitamin A, E, B_1, B_2 and B_6 in milk treated by microwave heating. (Sierra et al. 1999) found that continuous flow microwave treatment produce less destruction of vitamin B_1 in milk, which could attributed to the rapid temperature rise and the lack of hot surfaces in contact with milk in microwave system. However, other researchers have reported a significant loss of vitamin B_1 and found a thiamine of loss of over 50% in whole milk and 65% in skimmed milk after MP treatment at 80 ^{o}C for 4 min (Vidal-Valverde et al., 1993).

The impact of microwave heating on the main chemical changes, such as lactose isomerization, Maillard reaction and protein denaturation, taking place during process was also investigated, but there were also some disagreement. (Villamiel et al. 1996) reported that a rate enhancement of the chemical reactions occurred during microwave treatment in comparison with conventional heating. The difference was due to uneven heating of the milk in the microwave oven. Nevertheless, researchers found none of Maillard reaction products showed significant differences as between the microwave heating and conventional heating (Merbner et al., 1996), and low degree of whey protein denaturation was found after application of the continuous microwave treatment (Villamiel et al., 1996).

With regard to the sensorial changes in milk pasteurized by MP, it was been found that volatile composition was similar between MP-treated milk and conventional heating treated milk. Although no qualitative differences were found between microwave and conventional heated samples, when milk was heated in closed vessels some quantitative differences were found between the two heating systems, as well as during their storage (Valero et al., 1999). Furthermore, the sensory quality was the same for microwave and conventionally treated milk and no off-flavor were detected by sensory evaluation (Valero et al., 2000).

2.5. Micro Filtration Processing (MFP) of Milk

Micro filtration a novel membrane separation technology that uses micro-membrane with the pore sizes range from 0.1 - 10 μm to separate the molecules with different sizes. In recent years, the micro filtration processing techniques have been proposed for the reduction or elimination of microorganisms in the fluid milk products. The advantage of micro filtration is that it can remove microorganisms from the fluid milk without damaging the nutrients in the milk when compared with the conventional thermal inactivation of bacteria.

2.5.1. Effect of MFP on Microorganisms of Milk

The principle of the microfiltration applied in removing the microorganisms were illustrated in the Figure 2. The microorganisms often have a larger size than the pore of microfiltration membrane. For example, the sizes of (*Bacillus)* were from 0.5 – 30 μm, which often occur single, pairs and chains, resulting in the separation with other component of milk. According to the literatures, the microfiltration of milk reduced the (*B. cereus)* spore count by 99.95 - 99.98% and the total count by 99.99% (Kosikowski and Mistry, 1990; Olesen & Jensen, 1989). This significant spore reduction could not be obtained in the pasteurization processing. (Madec et al. 1992) has investigated that retention of (*Listeria)* and (*Salmonella)* cells in contaminating skim milk by tangential membrane microfiltration. They presented that the decimal reductions observed at 35 ^{o} C were close to 1.9 units for (*Listeria)* and 2.5 units for (*Salmonella)*. Moreover, unlike the thermal treatment, the reduction of bacteria was not influenced by contamination level (between 10^2 and 10^6 CFU/mL). An increase of microfiltration temperature could result in a significant increase of (*Salmonella)* retention (only 0.05% of the bacteria added were found in the retentate), but placed no effect on (*Listeria)* retention. Actually, the UTP device was introduced in the Bacto-catch process in order to produce ESL milks. It is possible to produce fluid milks having 30 cfu/mL mesophilic counts (compared with 900–3000 cfu/mL for conventional pasteurized milk ;(Saboya & Maubois, 2000), indicating the high efficiency of the microfiltration in removing the bacteria. The shelf life of the milk which has been processed by the microfiltration could extend 6-8 more days than the conventional pasteurized milk. There were also other reports which proposed 8-12 days extension for the shelf life of micro filtrated milk (Goff & Griffiths, 2006). Although there are many other effective non-thermal technologies, such as the bactofugation, the decimal reduction factor of microfiltration is much higher when compared to the bactofugation (Brans et al., 2004).

2.5.2. Effect of MFP nn Milk Quality

The microfiltration could not only exterminate the microorganisms in the milk but are also effective in maintaining the nutrient of the milk. Many researches have shown that microfiltration was not inducing significant changes in overall milk composition (Bindith, et al., 1996). (Hoffmann et al. 2006) focused their research on the processing of extending the shelf life of milk using microfiltration. They concluded that the microfiltration led only to a negligible change in the content of the main components of the ESL product when compared with the source milk. They also found that the minimum fat content is prescribed by law anyhow, and the total protein was only slightly decreased by microfiltration (0.02–0.03%); and the ratio of the protein fractions was unchanged within the accuracy of measurement. The same was valid for lactose and calcium. In addition, the shelf life of the ESL milk was distinctly prolonged than that of HTST-pasteurized milk without the significant changes of the sensory analysis for the micro filtrated milk. (Pafylias et al. 1996) has studied the microfiltration of milk with ceramic membranes. Their research concentrated on the nutrient and microbe changes of the milk by the microfiltration processing. The results showed that the protein, lactose, and minerals did not change significantly with a bacterial reduction of 4-5 log cycles. However, the microfiltration reduced the fat content in the milk because the diameter of some fat globules was larger than the pore sizes (James et al., 2003).

Although the microfiltration was proved to be an effective method for removal of the bacteria and maintain the nutrient in the milk, there are still many issues need has to be solved, such as high cost of the microfiltration membrane and the fouling of the membrane. In the future, the microfiltration may find greater application in removing bacteria from milk.

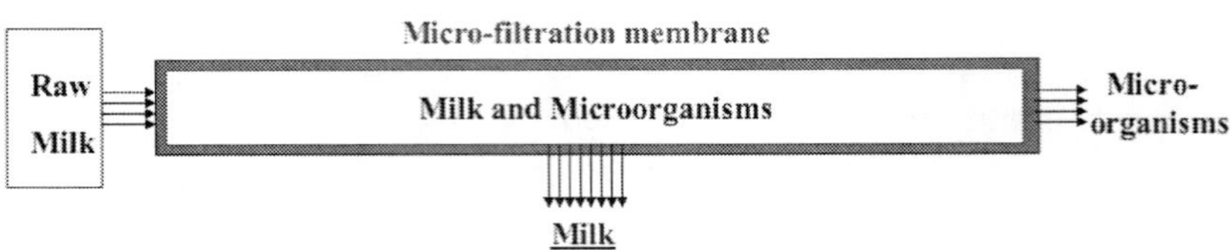

Figure 2. the principle of the Microfiltration processing of Milk.

REFERENCES

1. Conventional Pasteurization of Milk

Andersson I, Östea R, (1994) Nutritional quality of pasteurized milk. Vitamin B12, folate and ascorbic acid content during storage, *International Dairy Journal*, 4, 161-172.

Beddows C G, Blake C, (1982) the status of fluoride in bovine milk, II, The effect of various heat treatment processes, *J. Food Technol*, 63-70.

Boor K J, (2001) Fluid dairy product quality and safety: looking to the future, *J. Dairy Sci,* 84, 1–11

Bren L, (2004) *Got Milk, Make Sure its Pasteurized, FDA Consumer Magazine*, September-October 2004 Issue.

Canadian Food Inspection System - *Dairy Production and Processing Regulations* (Fourth Edition) - 2005

Čanigová M, Rajtarová K, Kakalej M, (2002) the influence of selected detergents on psychrotrophic microflora isolated from milk. Proceedings of lectures and posters ,Milk and milk products at the beginning of new millennium, *Hygiena Alimentorum*, 22, 54-58.

Champagne C P, Laing R R, Roy D, Mafu A A, (1994) *Psychrotrophs in dairy products : their effects and their control, Critical reviews in food science and nutrition*, 34, 1- 30.

Cifelli J C, Maples IS, Miller G, D, (2010) Pasteurization : Implications for Food Safety and Nutrition, *Nutrition Today*, 45, 207-213.

Cromie S J, (1991) Microbiological aspects of extended shelf life products, *Aust J Dairy Technol*, 46, 101–4

Cronjé M, (2003) *Production of Kepi Grains Using Pure Culture as Starters*. Master Thesis. Stellenbosch, South Africa: Stellenbosch University

Dairy Management Inc. (2003). Longer Shelf-Life Dairy-Based Beverages: Challenges and Opportunities. [Electronic version]. *Dairy Industry Technology Review*, December 2003.

Douglas S A, Gray M J, Crandall A D, Boor K J, (2000) Characterization of chocolate milk spoilage patterns, *J Food Prot*, 63, 516–521.

Dumalisile P, Witthuhn R C, Britz T J, (2005) Impact of different pasteurization temperatures on the survival of microbial contaminants isolated from pasteurized milk, *International Journal of Dairy Technology,* 58, 74-82.

Fox B A, Cameron A G, (1982) *Food Science—a Chemical Approach*, 4th edition, p 380. London: Hodder and Stoughton

Fromm H I, Boor K J, (2004) Characterization of Pasteurized Fluid Milk Shelf-life Attributes, *Journal of Food Science*, 69, 207-214.

Gandy A L, Schilling M W, Coggins P C, White C H, Yoon Y, Kamadia V V, (2008) The Effect of Pasteurization Temperature on Consumer Acceptability, Sensory Characteristics, Volatile Compound Composition, and Shelf Life of Fluid Milk, *Journal of Dairy Science*, 91, 1769-1777.

Lund D B J, (1982) Growth of thermo resistant streptococci and deposition of milk constituents on plates of heat exchangers during long operating times, *J. Food Protection,* 45(9), 806–812, 815.

Meer R R, Baker J, Bodyfelt F W, Griffiths M W, (1991) Psychrotrophic Bacillus spp. in Fluid Milk Products: A Review, *J. Food Protect*, 54, 969-979.

Miller G D, (2000) Handbook of Dairy Foods and Nutrition (2nd ed.). New York: CRC Press.

Petrus R R, Loiola C G, Oliveira C A F, (2010) Microbiological Shelf Life of Pasteurized Milk in Bottle and Pouch, *Journal of Food Science*, 75, 36-40.

Phillips J D, Griffiths M W, (1990) Pasteurized dairy products: The constraints imposed by environmental contamination, *Appl. Bacterial*, 61, 275-285.

Renner E, (1986) Nutritional aspects-Part I-Biochemical composition of pasteurized milk. *Bulletin of the International Dairy Federation*, No. 200, Chapter VII, pp. 27–29.

Wong N P, (Ed), (1999) *Fundamentals of Dairy Chemistry*. Gaithersburg: Aspen Publishers, Inc

1.1. HHP Processing of Milk

Chawla R, Patil G A, Singh A K (2011) High hydrostatic pressure technology in dairy processing: a review. *Journal of Food Science and Technology*, 48, 260-268.

Cheftel J C (1992). Effect of high hydrostatic pressure on food constituents -- an overview. In: Balny RH, Heremans H, Masson K (Eds) *High pressure and biotechnology*, Colloque INSERM. John Libbey & Co. Ltd, London

Considine T, Patel H A, Anema S G, Singh H, Creamer L K, (2007) Interactions of milk proteins during heat and high hydrostatic pressure treatments – A review, *Innovative Food Science and Emerging Technologies,* 8, 1-23.

Gaucheron F, Famelart M H, Mariette F, Raulot K, Michel F, Le Graet Y, (1997) Combined effects of temperature and high-pressure treatments on physicochemical characteristics of skim milk, *Food Chemistry*, 59, 439-447.

Hite B H, (1899) *The effect of pressure in the preservation of milk*, Bulletin of West Virginia University Agricultural Experiment Station, 58, 15-35.

Huppertz T, Fox P F, De Kruif K G, Kelly A L, (2006) High –Pressure-induced changes in bovine milk: a review. *International Journal of Dairy Technology*, 59, 58-66.

Huppertz T, Fox P F, de Kruif K G, Kelly A L, (2006a) High pressure-induced changes in bovine milk proteins: A review. *Biochimica et Biophysica Acta*, 1764, 593-598.

Huppertz T, Fox P F, Kelly A L, (2004) Effect of cycled and repeated high pressure treatment on casein micelles and whey proteins in bovine milk, *Milchwissenschaft* 59, 123-126.

Huppertz T, Kelly A L, Fox P F, (2006b) High pressure induced changes in ovine milk: effects on casein micelles and whey proteins. *Milchwissenschaft*, 61, 394-397.

Lopez-Fandino R, De la Fuente M A, Ramos M, Olano A, (1998) Distribution of minerals and proteins between the soluble and colloidal phases of pressurized milks from different species. *Journal of Dairy Research*, 65, 69–78.

Lopez-Fandino R, Olano A, (1998) Effects of high pressures combined with moderate temperatures on the rennet coagulation properties of milk, *International Dairy Journal*, 8, 623-627.

Ohmiya K, Kajino T, Shimizu S, Gekko K, (1989) Dissociation and reassociation of enzyme-treated caseins under high pressure, *Journal of Dairy Research*, 56, 435–442.

Payens T A J, Heremans K, (1969) Effect of pressure on the temperature-dependent association of β-casein, *Biopolymers*, 8, 335-345.

Rademacher B, Kessler H G, (1997) High pressure inactivation of microorganisms and enzymes in milk and milk products. In: Heremans K (Ed) *High pressure bioscience and biotechnology*. Leuven University Press, Leuven, pp 291–293.

Rendueles E, Omer M K, Alvseike O, Alonso-Calleja C, Capita R, Prieto M, (2011) Microbiological food safety assessment of high hydrostatic pressure processing: A review. *LWT-Food Science and Technology*, 44, 1251-1260.

Schrader K, Buchheim W, (1998) High-pressure effects on the colloidal calcium phosphate and the structural integrity of micellar casein in milk II. *Kinetics of the casein micelle disintegration and protein interactions in milk.* Kieler Milchwirtschaftliche Forschungsberichte, 50, 79-88.

Smelt J M, (1998) Recent advances in the microbiology of high pressure processing. *Trends Food Science and Technology,* 9, 152-158.

Vazquez-Landaverde R A, Torres J A, Qian M C, (2006) Effect of high-pressure-moderate-temperature processing on the volatile profile of milk. *Journal of Agricultural and Food Chemistry,* 54, 9184-9192.

Zobrist M R, Huppertz T, Uniacke T, Fox P F, Kelly A L (2005) High-pressure-induced changes in rennet-coagulation properties of bovine milk. *International Dairy Journal,* 15, 655-662.

1.2. PEF Processing of Milk

Barbosa- Gánovas G V, Góngora-Nieto M M, Pothakamury U R, Swanson B G, (1999) PEF-induced biological changes. In: *Preservation of Foods with Pulsed Electric Fields*. Pp 76-107. San Diego, CA: Academic Press.

Bendicho S, Barbosa- Gánovas G V, Martín O, (2002) Milk processing by high intensity pulsed electric fields, *Trends in Food Science & Technology*, 13, 195-204.

Damar S, Bozoğlu F, Hızal M, Bayındırlı A, (2002) Inactivation and injury of Escherichia coli O 157 : H 7 and Staphylococcus aureus by pulsed electric fields, *World Journal of Microbiology & Biotechnology*, 18, 1-6.

Fernández-Molina J J, Barbosa-Cánovas G V, Swanson G G, (2005) Skim milk processing by combining pulsed electric fields and thermal treatments, *Journal of Food Process Preservation,* 29, 291-306.

Fernández-Molina J J, Fernández-Gutiérrez S A, Altunakar B, Bermúdez-Aguirre D, Swanson G G, Barbosa-Cánovas G V, (2005) The combined effect of pulsed electric fields and conventional heating on the microbial quality and shelf life of skim milk. *Journal of Food Process Preservation*, 29, 390-406.

Floury J, Grosset N, Leconte N, Pasco M, Madec M, Jeantet R, (2006) Continuous raw skim milk processing by pulsed electric field at non-lethal temperature: effect on microbial inactivation and functional properties, *Dairy Science and Technology*, 86, 43-57.

Garcia-Amezquita L E, Primo-Mora A R, Barbosa-Cánovas G V, Sepulveda D R, (2009) Effect of nonthermal technologies on the native size distribution of fat globules in bovine cheese-making milk, *Innovative Food Science and Emerging Technologies*, 10, 491-494.

Lindgren M, Aronsson K, Galt S, Ohlsson T, (2002) Simulation of the temperature increase in pulsed electric field (PEF) continuous flow treatment chambers. *Innovative Food Science and Emerging Technologies,* 3, 233-245.

Martín O, Quin B L, Chang F J, Barbosa-Cánovas G V, Swanson B G, (1997) Inactivation of Escherichia coli in skim milk by high intensity pulsed electric fields. *Journal of Food Process Engineering*, 20, 317-336.

Michalac S, Alvarez I, Zhang QH (2003) Inactivation of selected microorganisms and properties of pulsed electric field processed milk. *Journal of Food Process Preservation,* 27, 137-151.

Otunola A, El-Hag A, Jayaram S H, Anderson W A, (2008) Effectiveness of pulsed electric fields in controlling microbial growth in milk. *International Journal of Food Engineering,* 4, 1-14.

Qin B, Pothakamury U R, Barbosa-Gánovas G V, Swanson B G, Peleg M, (1996) Nothermal pasteurization of liquid foods using high-intensity pulsed electric fields, *Critical Reviews in Food Science and Nutrition*, 36, 603-627.

Raso J, Góngora M M, Calderón M L, Barbosa-Cánovas G V, Swanson B G, (1999) Resistant micro-organisms to high intensity pulsed electric field pasteurization of raw skim milk. *IFT annual meeting technical program. IFT annual meeting technical program*. Chicago, Illinois: Institute of Food Technologists.

Saulis G, (2010) Electroporation of cell membranes: the fundamental effects of pulsed electric fields in food processing, *Food Engineer Research*, 2, 52-73.

Smith K, Mittal G S, Griffiths M W, (2002) Pasteurization of milk using pulsed electrical field and antimicrobials, *Journal of Food Science*, 6, 2304-2308.

Sobrino-López A, Matín-Belloso O, (2010) Review: Potential of high-intensity pulsed electric field technology for milk processing, *Food Engineering Reviews*, 2, 17-27.

Zhang S, Yang R, Zhao W, Liang Q, Zhang Z, (2011) The first observation of radical species generated under pulsed electric fields processing, *LWT-Food Science and Technology*, 44, 1233-1235.

1.3. UP Processing of Milk

Ashokkumar M, Bhaskaracharya R, Kentish S, Lee J, Palmer M, Zisu B, (2010) The ultrasonic processing of dairy products – An overview, *Dairy Science and Technology*, 90, 147-168.

Bermúdez-Aguirre D, Corradini M G, Mawson R, Barbosa-Cánovas G V, (2009) Modeling the inactivation of Listeria innocua in raw whole milk treated under thermo-sonication, *Innovative Food Science and Emerging Technologies*, 10, 172-78.

Cameron M, McMaster L D, Britz T J, (2009) Impact of ultrasound on dairy spoilage microbes and milk components, *Dairy Science and Technology*, 89, 83-98.

Chouliara E, Georgogianni K G, Kanellopoulou N, Kontominas M G, (2010) Effect of ultrasonication on microbiological, chemical and sensory properties of raw, thermized and pasteurized milk, *International Dairy Journal*, 20, 307-313.

Garcia M A, Burgos J, Sanz B, Ordonez J A, (1989) Effect of heat and ultrasonic waves on the survival of two strains of Bacillus subtilis, *Journal of Applied Bacteriol*, 67, 619-628.

Gera N, Doores S, (2011) Kinetics and mechanism of bacterial inactivation by ultrasound waves and sonoprotective effect of milk components, *Journal of Food Science*, 76, 111-119.

Harvey E N, Loomis A L, (1929) The destruction of luminous bacteria by high frequency sound waves, *Journal of Bacteriol*, 17, 373-376.

Hoover U, (2000) Kinetics of microbial inactivation for alternative food processing technologies: ultrasound. *Journal of Food Science, Supplement*, 93-95.

Kinsloe H, Ackerman E, Reid J J, (1954) Exposure of microorganisms to measured sound fields, *Journal of Bacteriology*, 68, 373-380.

Mason T J, Lorimer J P, (2002) Introduction to applied ultrasonics. In: *Applied Sonochemistry: The uses of power ultrasound in chemistry and processing*. Pp. 1-24. Weinheim, Germany: Wiley VCH.

Piyasena P, Mohareb E, Mckellar R C, (2003) Inactivation of microbes using ultrasound ''a review''. *International Journal of Food Microbiology*, 87, 207-216.

Riener J, Noci F, Cronin D A, Morgan D J, Lyng J G, (2009) Characterisation of volatile compounds generated in milk by high intensity ultrasound, *International Dairy Journal*, 19, 269-272.

Riesz P, Kondo T, (1992). Free radical formation induced by ultrasound and its biological implications. *Free Radical Biology and Medicine*, 13, 247-270.

Sala F J, Burgos J, Condón S, Lopez P, Raso J, (1995) Effect of heat and ultrasound on microorganisms and enzymes. In: *New methods of food preservation* (Gould G W). Pp. 176-204. London: Blackie Academic & Professional.

Taylor M J, Richardson T, (1980) Antioxidant activity of skim milk: effect of sonication, *Journal of Dairy Science*, 63, 1938-1942.

1.4. MP Processing of Milk

Brown G H, Morrison W C, (1954) An extrapolation of the effects of strong radiofrequency fields on microorganisms in aqueous solutions, *Food Technology*, 8, 361-366.

Chiu C P, Tateishi K, Kosikowski F V, Armbruster G, (1984) Microwave treatment of pasteurized milk, *Journal of Microwave power*, 19, 269-272.

Flemming H, (1944) Effect of high-frequency fields on microorganisms. *Electrical Engineering*, 63, 18-22.

Hamid M A K, Boulanger R J, Tong S C, Gallup R A, Pereira R R, (1969) Microwave pasteurization raw milk, *Journal of Microwave Power*, 4, 272-275.

Knutson K M, Marth E H, Wagner M K, (1988) Use of microwave ovens to pasteurize milk, *Journal of Food Protection*, 51, 715-719.

Lechowich R V, Beuchat L R, Fox K I, Webster F H, (1969) Procedure for evaluating the effects of 2,450-megahertz microwaves upon Streptococcus faecalis and Saccharomyces cerevisiae, *Applied Microbiology*, 17, 106-110.

Merbner K, Erbersdobler H F, (1996) Maillard reaction in microwave cooking: comparison of early maillard products in conventionally and microwave-heated milk, *Journal of the Science of Food and Agriculture,* 70, 307-310.

Merin U, Rosenthal I, (1984) Pasteurization of milk by microwave irradiation. *Michwissenschaft*, 39, 643-644.

Sierra I, Vidal-Valverde C, Olano A, (1999) The effects of continuous flow microwave treatment and conventional heating on the nutritional value of milk as shown by influence on vitamin B1 retention, *European Food Research and Technology*, 209, 352-354.

Siever R, Eberhard P, Fuchs D, Gallmann P U, Strahm W, (1996) Effect of microwave heating on vitamins A, E, B1, B2 and B6 in milk. *Journal of Dairy Research*, 63, 169-172.

Valero E, Sanz J, Martínez-Castro I, (1999) Volatile components in microwave- and conventionally-heated milk*, Food Chemistry*, 66, 333-338.

Valero E, Villamiel M, Sanz J, Martínez-Castro I, (2000) Chemical and sensorial changes in milk pasteurized by microwave and conventional systems during cold storage, *Food Chemistry*, 70,77-81.

Vela G R, Wu J F, (1979) Mechanism of lethal action of 2,450-MHz radiation on microorganisms. *Applied and Environmental Microbiology*, 37, 550-553.

Vidal-Valverde C, Redondo P, (1993) Effect of microwave heating on the thiamine content of cow's milk. *Journal of Dairy Research*, 60, 259-262.

Villamiel M, Corzo N, Martínez-Castro I, Olano A, (1996) Chemical changes during microwave treatment of milk, *Food Chemistry*, 56, 385-388.

Villamiel M, López-Fandiño R, Corzo N, Martínez-Castro I, Olano A, (1996) *Effects of continuous flow microwave treatment on chemical and microbiological characteristics of milk*, Z Lebensm Unters Forsch, 202, 15-18.

Young G S, Jolly P G, (1990) Microwaves: the potential for use in dairy, *Australian Journal Dairy Technology*, 45, 34-37.

1.5. MFP Processing of Milk

Bindith O, Cordier J L, Jost R, (1996) Cross- flow microfiltration of skim milk: Germ reduction and effect on alkaline phosphatase and serum proteins. In *Heat treatments and alternative methods: Bulletin 9602* (pp. 222–231). Brussels, Belgium: International Dairy Federation.

Brans G, Schroën C G P H, van der Sman R GM, Boom R M, (2004) Membrane fractionation of milk: State of the art and challenges, *J. Member. Sci.*, 243, 263-272.

Goff H D, Griffiths M W, (2006) Major advances in fresh milk and milk products: Fluid milk products and dairy desserts, *Journal of Dairy Science*, 89,1163–1173.

Hoffmann W, Kiesner C, Clawin-rädecker I, Martin D, Einhoff K, Lorenzen P C, Meisel H, Hammer P, Suhren G, Teufel P, (2006) Processing of extended shelf life milk using microfiltration, *International Journal of Dairy Technology*, 59, 229-235.

James B J, Jing Y, Chen X D, (2003) Membrane fouling during filtration of milk—a micro structural study, *Journal of Food Engineering*, 60, 431-437.

Kosikowski F V, Mistry V V, (1990) Microfiltration, ultrafiltration, and centrifugation separation and sterilization processes for improving milk and cheese quality, *J. Dairy Sci.*, 73, 1411-1419.

Madec M N, Mejean S, Maubois J L, (1992) Retention of Listeria and Salmonella cells contaminating skim milk by tangential membrane microfiltration, Lait, 72, 327–332.

Olesen N, Jensen F, (1989) Microfiltration: The influence of operation parameters on the process, *Milchwissenschaft*, 44, 476-479.

Pafylias I, Cheryan M, Mehaiab M A, Saglam N, (1996) Microfiltration of milk with ceramic membranes, *Food Research International*, 29, 141-146.

Saboya L V, Maubois J L, (2000) Current developments of microfiltration technology in the dairy industry, *Lait*, 80, 541–553.

In: Raw Milk
Editors: J. Momani, et al
ISBN: 978-1-61470-641-0

Chapter 7

Controlled Atmosphere-Based Improved Storage of Cold Raw Milk: Potential of N_2 Gas

Patricia Munsch-Alatossava** and Tapani Alatossava
Department of Food and Environmental Sciences,
Division of Food Technology,
FIN-00014 University of Helsinki, Finland

Abstract

On one hand, according to FAO about 80% of the milk consumed worldwide is mostly obtained out of standards; in developed countries on the other hand an effective cold chain selects for spoiling bacteria that inflict significant losses to the dairy industry. Most studies, that concern modified or controlled atmospheres applied to bovine raw milk, were mostly based on CO_2 treatments, or for a few on mixtures of CO_2 and N_2 gases; a commonly accepted thought is that antimicrobial effects are associated with the application of CO_2, whereas N_2 has been employed as an inert gas component. Some recent studies, performed with an open system, based on a constant flushing of N_2 gas through the headspace of a vessel, at laboratory or at pilot scale suggest that bacterial growth could be substantially reduced by flushing pure N_2 gas alone into raw milk, with significant effects on mesophilic and psychrotrophic aerobes, but

* Emails: patricia.munsch@helsinki.fi; tapani.alatossava@helsinki.fi

also on some other bacterial groups, without favouring the growth of anaerobes. One major observation was that phospholipases producers among them *Bacillus cereus* could be excluded at laboratory scale by the N_2 gas-based flushing; the inhibitory effect was also noticeable to some extend at pilot scale. Possible antimicrobial mechanisms underlying the use of N_2 gas, as well as the potential of controlled atmospheres-based treatments of raw milk will be discussed.

1. Supply and Quality of Food in a Climate Changed World

Continuing population and consumption growth will mean that the global demand for food will increase for at least another 40 years (Godfray *et al.* 2010). Recent studies estimate the need from 70 to 100% more food by 2050; this should be achieved by the production of food considering the present environmental constrains as finite resources and ongoing climate changes. All steps from production, storage, processing, until distribution are consequently under challenge. To overcome, the past drifts, food production systems and the food chain must become fully sustainable without neglecting safety aspects. The challenge of feeding 9 billion people by roughly the middle of this century (Godfray *et al.* 2010) requests different measures: among them reducing waste. Roughly 30 to 40% of food in both the developed and developing worlds is lost to waste. In developing countries losses are mainly due to the absence of food-chain infrastructure, of lack of investment in storage technologies (cold storage for example); immediate selling is requested (subsistence farming). In the developed world the losses also raise for different reasons.

Considering food borne diseases, "the challenges of 20 years ago still persist while new ones continue to emerge"(Newell *et al.* 2010). Many factors along the cold chain affect the microbiological safety of food, and although food production practices change, the well know food borne pathogens such as *Salmonella spp.* and *Escherichia coli* showed remarkable ability to exploit novel opportunities and generate new challenges such as antibiotic resistance (Newell *et al.* 2010). The experience from the last 20 years indicate for many countries, including in Europe where the food production is qualified as high-tech and has never been more stringently controlled, consumers still suffer from food borne diseases, the major bacterial pathogens still constitute serious threats, by evolving when facing new challenges, occupying new niches, and displaying new virulence properties (Jakobsen 2010, Newell *et al.* 2010).

2. Spoilage of Food

Worldwide food spoilage constitutes an enormous economic problem. It is estimated that one-fourth of the world's food supply is lost through microbial activity alone (Huis int'Veld 1996). Food spoilage may be considered as any change that renders a product unacceptable for human consumption from a sensory point of view (Hayes 1985, Gram *et al.* 2002), and the consequence of a complex event in which a combination of microbial and (bio)chemical activities may interact (Huis int'Veld 1996, Gram 2002). Microbial spoilage is by far the most common cause of spoilage and may manifest itself as visible growth (slime production, apparition of colonies), as textural changes (degradation of polymers), or as off-odours and off-flavours (Gram *et al.* 2002). Refrigeration stops or reduces the rate at which changes occur in food; the thought that food properly refrigerated would remain safe was persistent until several pathogens like *Aeromonas hydrophila*, *Listeria* spp, some strains of *Bacillus cereus*, enteropathogenic *E. coli* or non proteolytic strains of *Clostridium botulinum* that can grow at refrigeration temperatures arose (Marth 1998). The safety and quality of many foods rely on refrigeration, which if extended would permit foods to be distributed to an increasing urbanised world; noteworthy less than 10% of perishable foods are in fact currently refrigerated, though a more generalised cold storage would have implications on greenhouse gas emissions (Coulomb 2008, James and James 2010). Already 15% of the electricity consumed worldwide is used for refrigeration; if no alternative systems are developed in order to extend and improve the cold chain this leads inescapably to higher energy consumption with a rise in ambient temperature (James and James 2010).

3. Microbes in Milk

3a. Microbial Diversity and Milk Quality

Milk as a highly nutritious food constitutes also an excellent growth medium for a wide range of microorganisms (Table I); due to multiple contamination sources, many different types of bacteria are present in raw milk irrespective of their growth optimum; the types and amounts reflect season variation and include bacteria with human pathogenic potential (*Listeria monocytogenes*, *Salmonella* spp, *Staphylococcus aureus*, *Mycobacterium tuberculosis*), psychrotrophic bacteria belonging to the genera *Pseudomonas, Enterobacter, Flavobacterium, Klebsiella, Aeromonas, Acinetobacter, Alcaligenes, Achromobacter, Serratia, Vibrio*; certain genera host species that are both psychrotrophic and thermoduric (*Bacillus, Clostridium, Microbacterium*) (Cousin 1982, Hayes and Boor 2001, Chambers 2002).

Table I. Raw milk microflora (modified from Franck and Hassan 2002)

Microorganisms	Incidence
Yeast, Moulds	< 10%
Micrococcus, Staphylococcus	30-99%
Streptococcus, Lactococcus	0-50%
Lactobacillus, Corynebacterium, Microbacterium	< 10%
Gram negative bacteria: *Pseudomonas, E.coli, Alcaligenes, Acinetobacter*	< 10%
Gram positive bacteria: spore formers:*Bacillus, Clostridium*	< 10%

The storage temperature and the elapsed time after raw milk´s collection both determine the evolution of the microflora. When milk is stored below 4°C, bacterial multiplication is delayed by 24h at least; however, shortly after 48h, the low temperature does not prevent bacterial growth (the so-called critical age is reached). In developed countries, the indicator for monitoring the sanitary conditions of raw milk is the "total" bacterial count or SPC (standard plate count): the standard for raw milk Grade A or 1 relies on an SPC value below 1.0×10^5 CFU/ml (EC legislation 2001, Chambers 2002). Cooling of raw milk below 6°C, typically at 3 to 4°C at the farm tank following milking, followed by storage at low temperatures (below 6°C) during transportation to the dairy plant aims to ensure the quality of raw milk until its entrance to the different dairy processes which often include a heat-treatment step (typically pasteurisation or UHT treatment) as a critical point (CCP) for HACCP-based food safety management. The counts should not exceed 3.10^5 CFU/ml before the milk is processed. According to FAO, over 80% of the milk consumed in developing countries (200 billion litres annually) is handled by informal market traders, with inadequate regulation: smallholder farmers are predominant, no cold chain exists, dairy farming is not that advanced technologically, and milk may be travelling via public transportations, by bike or by foot to collection centres; not many options are yet available to fully exploit the opportunities for livestock development, to alleviate poverty while improving safety and minimizing waste (FAO 2009, Kisaalita 2010).

3b. Psychrotrophs as Spoiling Agents

The cold storage of foodstuffs has selected for a category of microorganisms comprising bacteria, yeasts, moulds which can grow at temperatures below 7°C, with an optimal and maximal growth at temperatures ranging between 25-30°C, and 30-35°C respectively. The majority of the bacterial genera that constitute the psychrotrophic community are Gram negative representatives (Jay *et al.* 2005). In milk and dairy products, most psychrotrophic bacteria usually come from soil, water, and vegetation; the amounts are generally lower at farm milk tank compared to bulk tanks; the occurrence of psychrotrophs reflect the sanitary conditions, the age of the raw milk.

The cold storage together with the chemical composition of the milk itself favours the growth of psychrotrophic bacteria for which the perfection in adaptation is reached by their production of exoenzymes (like proteases, lipases, phospholipases) that withstand the classical heat treatments of the milk (Fox et al. 1976, Cousin 1982, Hayes and Boor 2001, Chambers 2002). Lipolytic and proteolytic activity is the apanage of many psychrotrophs which alter the different milk components and induce rancid flavours/odours for milk or bitter flavour/ coagulation of dairy products, and consequently inflict significant qualitative and quantitative losses to the dairy industry. The potential to degrade both raw and processed milk components (by thermoduric genera), may explain why raw milk psychrotrophs are mainly considered due to their spoilage features as benign bacteria, to the exception of the human pathogens *Bacillus cereus* (toxin producing strains) or *Listeria monocytogenes*. Recently and worldwide, Gram negative bacteria are under higher scrutiny since many genera host species considered as human opportunistic pathogens, which carry antibiotic multiresistant traits (McGowan 2006). When characterizing some raw milk spoiling gram negative-psychrotrophs, we could observe that isolates carried antibiotic resistance

(AR) features that seemed to increase along the cold chain of milk storage and transportation (Munsch-Alatossava and Alatossava 2006, 2007).

4. Modified and Controlled Atmospheres

4.1. History

Food storage is an important development for food production, sedentism, farming, and represents a major evolutionary threshold for human civilization (Kuit and Finlayson 2009). Recent excavations at Dhra′near the Dead Sea in Jordan provide strong evidence for sophisticated purpose-built granaries in a predomestication context -11300-11175 cal B.P ; suspended floors allowing air circulation bring evidence for food storage at Pre-Pottery Neolithic Age (Kuit and Finlayson 2009). A reasonable guess suggests that grains were stored "all that food shall be for store "during the 7 good years to survive the 7 years of famine (Genesis Chap 41/ 36). Early 19th century, botanists and physiologists started to investigate the effects of manipulating the composition of the atmosphere on the ripening of fruits; Bérard (1821) observed that fruits in an environment deprived of O_2 retained their original appearance but lost their ripening ability if kept too long. In 1877, Pasteur and Joubert reported that CO_2 can kill *Bacillus anthracis*. Still prior to 1900, food habits were adjusted to the availability of foods; in most climates this has been very greatly affected by the facilities to preserve foods during seasonal or famine periods (Woolrich 1944). The first practical use of modified atmospheres (MA), based on elevated levels of CO_2, aimed to preserve fresh meat carcasses on their way from New Zealand and Australia to Great Britain in the 1930s (Silliker and Wolfe 1980). Food preservation relies on heating, chilling, freezing, drying, salting, smoking removing of O_2 … applied first rather empirically; the use of multiple and sequential preservation factors so–called hurdles constitutes nowadays the bases of the hurdle technology which aims to improve the food ′s quality and safety throughout the different processing steps (Leistner and Good 2005). The increased need for fresher and safer ready- to-eat- products promoted the development of MA, and CA (Controlled Atmospheres) based extension of storage life of foods, which is a late 20th century application (Welsh and Mitchell 2000): according to Ben Yoshua et al. (2005), the understanding of the state of art of MA applications relies on thousands of years of practices and on more recent scientific and technological progresses. The precision of the control of partial gas pressures distinguishes

MA and CA: a single component of the atmosphere is modified for MA (which may passively establish), whereas in CA (actively installed, like by flushing), a higher degree of control is applied; an active technological control imposes constraints which are maintained by monitoring the requested adjustments (Welsh and Mitchell 2000, Raghavan et al. 2005). Improved storage technology based on MA and CA account among the innovative processing technologies, that led to numerous industrial applications considering whether MA packaging, or CA storage (Ben-Yoshua et al. 2005, O´Beirne 2010).

4.2. Principles and Applications

The atmosphere that overhangs earth has an approximate composition of 79% N_2, 21% O_2 and 0.04% CO_2. Although a wide range of gases has been considered such as ozone, argon, carbon monoxide, sulphur dioxide, most applications of modified atmospheres are based on the three main natural gases present in air, used as a single or as a combination of two, and at different levels as in the air. The applied treatments, usually based on a reduction of the O_2 level and a concomitant addition of CO_2 (from the order of parts per million up to 100%) or CO (less often) retard metabolic activities, oxidative reactions and inhibit the growth of spoiling or pathogenic bacteria. Numerous advantages are highlighted when CA or MA are applied to fruits and vegetables, to cereals and oilseeds to preserve grain from pests (Mazza and Jayas 2001, Ben Yoshua et al. (2005)). Many studies evaluated MA based treatments for fish and meat for which the 1st commercial application of MAP (Modified atmosphere packaging, MAP) was reported in 1979 when Marx and Spencer introduced MAP meat (Philipps 1996). MAP is very widely used to extend the shelf life of various foodstuffs including fresh chilled products, cooked perishable foods, long-life products (O´Beirne 2010). The MAP based extension of storage life consists in flushing a package of food with gases just before sealing it. MAP applied to dairy food products takes use of CO_2 and inert N_2: both gases are introduced directly into liquids or semi liquid foods (like milk, yoghurt, sour cream, ice cream and cottage cheese) (Alvarez and Ji 2003). MAP based on gas ratios of 50:50 or 40:60 of CO_2 and N_2 respectively were the most effective to control the growth of different bacterial groups (mesophiles, psychrotrophs, enterobacteria) in cameros cheeses (Gonzalez-Fandos et al 2000).

4.3. Carbon Dioxide (CO_2) and/or Nitrogen (N_2) Based MA and CA

4.3.1. Carbon Dioxide (CO_2)

CO_2 can either stimulate or inhibit the growth of microorganisms (Valley and Rettger 1927); if all microorganisms require a certain level of CO_2 in their metabolism, the so-called capnophiles grow better in the presence of a higher CO_2 tension than the level normally present in the atmosphere. CO_2 is responsible for the bacteriostatic effect seen on microorganisms grown in MA environments and constitutes the major anti-microbial factor of modified atmospheres. At high levels of CO_2, the microbial growth is reduced and the effect increases when the storage temperature decreases (Gill and Tan 1979). The use of CO_2 is usually associated with a drop of the pH. The antimicrobial properties depend on the type of food, the temperature of incubation, the gas concentration, the load of initial bacterial population and the microorganisms types. Aerobic gram negative bacteria are relatively sensitive to CO_2 contrarily to LAB which are quite resistant (Chen and Hotchkiss 1991, Farber 1991, Gorris and Peppenlenbos 2008, O´Beirne 2010). MAP-based on CO_2 and applied to fresh fruits and vegetables revealed that moulds were rather sensitive, yeasts more resistant, whereas *Pseudomonas*, *Micrococcus* and *Bacillus* were inhibited by CO_2; facultative anaerobes like *E.coli* were less affected by CO_2 but more sensitive to the level of O_2. The mode of action of CO_2 is not yet fully understood (Gorris and Peppenlenbos 2008, O´Beirne 2010) although the effect seems to be pleiotrophic; dissolved CO_2 inhibits bacterial growth in raw milk by affecting the three growth phases (lag, exponential and stationary phase), the maximum growth rate, and the maximum populations densities (King and Mabbitt 1982, Roberts and Torrey 1988, Farber 1991, Martin et al. 2003, Werner and Hotchkiss 2006). The overall inhibition was greater on gram-negative compared to gram-positive bacteria (Martin et al. 2003). About two decades ago, high pressure carbon dioxide (HPCD) inactivation of microorganism in foods was proposed to overcome the drawbacks of loss of tastes or flavours when foodstuffs were heat treated; several evidences and hypothesis are proposed to explain the antimicrobial effect of pressurised CO_2, although the mechanism is also not fully understood (Hong and Pyun 2001, Garcia-Gonzalez et al. 2007).

Some of the known effects induced by CO_2:

a) Changes in intracellular pH

Since CO_2 is highly soluble in both aqueous solutions and lipids, CO_2 can easily diffuse in an out of cells. The carbonic anhydrase enzyme catalyses the reversible hydratation of CO_2 into carbonate: $CO_2 + H_2O \leftrightarrow HCO_3^- + H^+$. The entrance of CO_2 leads to a pH drop, or acidification which affects the metabolic activities within the cell. The decrease in pH is amplified at lower temperatures when the gas solubility is higher (Wolfe 1980, Daniels et al. 1985, O'Beirne 2010).

b) Alteration of microbial protein and enzyme structure and function

King and Nagel (1975) observed that if CO_2 exceeds 50% certain exoenzymes are not expressed; Mitz (1979) reported conformational changes in enzymes in the presence of elevated CO_2, and according to Mac Mahon (2000), 50% of CO_2 enable the growth of *A. hydrophila* but both proteinase and haemolysin were not expressed. In the presence of CO_2, the solubility of the enzymes are modified following conformational changes leading to the inactivation of enzymes (Mitz 1979, Mac Mahon 2000).

c) Alteration of membrane structure and function

Dissolved CO_2 and the ions HCO_3^- altered the structure of bacterial cell membranes: HCO_3^- increases the hydratation of membranes contrarily to CO_2 that dehydrated membranes, with consequences on the export of enzymes, substrate uptake (King and Nagel 1967, Daniels 1985, Farber 1991).

d) Gene expression and metabolic regulation

When the concentration is sufficiently high, CO_2 may act as a metabolic regulator; elevated CO_2 concentrations may inhibit decarboxylation reactions in which CO_2 is released by feedback mechanisms (Dixon and Kell 1989). CO_2 regulates gene expression across a wide range of microorganisms including fungi, photosynthetic bacteria (like Cyanobacteria), as well as non photosynthetic bacteria (Stretton and Goodman 1998). In the presence of CO_2, putative virulence determinants of *Borrelia burgdorferi* are regulated at the transcriptional level: the bacterium alters its gene expression and antigenic profile (Hyde et al. 2007).

4.3.2. Nitrogen (N_2)

Rutherford discovered nitrogen in 1772 as another air component, Lavoisier recognized it as a simple element and named it azote (without life, as contrarily to O_2, it does not support breathing). Nitrogen is considered as chemically benign, inert, odourless and tasteless (Farber 1991, Philipps 1996,

Theriault et al. 2004) and is poorly soluble in water. Liquid N_2 is the most used cryogenic fluid to chill, freeze food products; N_2 gas enters into numerous applications like in the manufacture of stainless steel, the production of electronic parts like diodes or transistors; it is used for inerting (Theriault et al. 2004), for protection of historical documents (to avoid decay of paper and ink) drying or lyophilisation until the preservation of bulk or packaged foodstuffs… When N_2 replaces O_2 in MAP products, it delays oxidative rancidity and inhibits the growth of aerobic microorganisms (Farber 1991, Philipps 1996); different MAP food products are kept under mixtures of CO_2 and N_2 based atmospheres; for example mixtures of 0-70% CO_2 and 0-30% N_2 served to preserve cheeses (Farber 1991, Philipps 1996); N_2 prevents pack collapse which may occur if CO_2 is used in high contents (Farber 1991). N_2 is used as a filter gas because of its low solubility in water and lipid, as compared to CO_2 (Philipps 1996). The greatest hazard of N_2 is due to its asphyxiation properties when the percent of O_2 entering the lungs is too low to maintain essential levels of O_2 in the blood, and consequently endangers life itself (Weller 1959). The entrance and filling of the intramitochondrial space by N_2 blocks the uptake of O_2 and may lead to anaerobic metabolism, acidosis and cell death: in the case of cerebrovascular accidents or myocardial infarctions, nitrogen toxicity becomes a problem when the blood flow through organs is blocked; when the O_2 is exhausted in the blood flow compromised region, the mitochondrial membrane looses its integrity, and N_2 leaks into the mitochondria and further blocks the entrance of O_2 (Van Deripe 2010).

4.4. Control of Microorganisms Present in Raw Milk

In developed countries, the control of bacterial growth (mainly mesophiles and thermophiles) implies rapid cooling of raw milk below 6°C. Heat treatments (pasteurisation, ultra pasteurisation, and UHT) play a critical role in further controlling the different bacterial communities by achieving reduction of bacterial numbers until microbial sterility depending on the efficiency of heat-treatments. Bactofugation, a centrifugation- based method aims to remove bacterial spores; clarification, which relies on a difference of relative densities of bacterial cells and other foreign particles, separates milk components from somatic cells and other unwished particles. Microfiltration and ultra filtration can remove most of the bacteria (>99.9 % of vegetative and spore cells) (Hayes and Boor 2001, Chambers 2002). All previous listed methods present advantages and limitations, and mainly could not be applied worldwide wherever needed.

4.5. CO_2 and N_2 Gases Applied to Milk

Numerous studies (some are listed in Table II) reported an extension of shelf life of milk after the addition of carbon dioxide gas (CO_2) (King and Mabbitt 1982, Hotchkiss and Lee 1996, Ruas-Madiedo *et al.* 1996, Martin et al. 2003, Rajagopal et al. 2005, Dechemi et al 2005). For example, Ma et al. (2003) reported a decreased proteolysis following the addition of CO_2 to raw milk; less microbial proteases were produced due to a lower microbial growth; the pH drop was proposed to also alter the action of endogenous protease activity; the effect of CO_2 on lipolysis was mostly due to a reduced microbial growth: with 1500 ppm dissolved CO_2, the milk could be stored for 14d at 4°C with counts lower than 3.10^5CFU/ml. The efficiency highlighted by many studies (Table II) is indisputable, despite some disadvantages like a modification of the sensory properties, or the promotion of acidification of raw milk if the CO_2 is not eliminated prior to further processing of the milk.

Nitrogen (N_2) considered as an inert gas, has some potential to overcome the disadvantages of CO_2. Two studies investigated the treatment of raw milk with nitrogen gas (N_2) applied, to a close system (that did not enable gas exchanges between the flask containing the milk and the environment) (Murray *et al* 1983, Dechemi *et al.* 2005). Since raw milk tanks are open systems that allow gas balance between the headspace of the tank and the external environment, we investigated the application of a pure N_2 (99.999 %) gas flow-through system (an open system) to raw milk at laboratory scale (120 mL raw milk): like with CO_2, the inhibitory effect on certain spoiling bacterial groups was also evident: the system was of interest in a temperature range of 6°C to 12°C (Tables III and IV); the treatments do not induce acidification of the treated milks; at 12°C, the bacterial growth could be halted for 48h; surprisingly was noticed that phospholipases (PLs) producing bacteria were "sooner or later" excluded in raw milk at laboratory scale (Munsch-Alatossava *et al.* 2010 a,b: Table IV).

Table II. Modified and controlled atmospheres applied to raw milk

Gas	System applied /CO_2 concentrations	Volume of treated milk	T in °C/time	Beneficial effects	References
CO2	10-30mM	100 mL	4,7, 10°C	inhibition of bacterial growth increases when CO_2 increases, and when the storage temperature decreases	King and Mabbitt 1982
	food grade CO_2 bubbled in milk until the pH was lowered to 6.0 and 6.2), followed by degasification	40L/pilot scale	4°C /4d	inhibition of coliforms, of proteolytic and lipolytic psychrotrophs	Ruas-Madiedo et al. 1996
	sparging milk with 11.9 mM CO_2		6.1°C /35d	addition of CO_2 had no stimulatory or inhibitory effect on the initiation of germination, or on the outgrowth of *Bacillus cereus* spores	Werner and Hotchkiss 2002
	1500ppm (pH 6.2 at 4°C)		4°C/21d	delayed bacterial growth; decreased proteolysis and lipolysis until 14d	Ma et al. 2003
	CO_2 bubbled in milk (0.6 to 61.4 mM)		15°C/4-5d	CO_2 affects the different growing phases in a gram-dependant response (evaluated by conductance)	Martin et al. 2003

	Pressurised CO_2 at 10.3 – 48.3 MPa; and CO_2 concentrations of 0-132g CO_2 /Kg of milk		30-35°C	higher total microbial lethality with supercritical CO_2 and with 132g/Kg of milk at 35°C	Werner and Hotchkiss 2006
	100 mL/min for 25 min ; CLOSED SYSTEM	250 mL	7°C storage for 10d	inhibition of psychrotrophs (78%), Pseudomonadacese (75%), Enterobacteria (99%) and of 100% of protease and lipase activity	Dechemi et al. 2005
$CO_2 + N_2$ (25 to 75%)	99,95 % purity: 100 mL/min for 25 min ; CLOSED SYSTEM	250 mL	7°C storage for 10d	Mixture of 50% CO_2 and N_2: highest inhibitory effect on psychrotrophs (98 %), Pseudomonadacese (98.9%), Enterobacteria (99.5%), Lactic acid bacteria (98.4%); 100% and 98% of lipase and protease activity respectively	Dechemi et al. 2005
N2	100mL/min (USP grade): CLOSED SYSTEM	1L	4°C/13 d	longer lag phase and slower growth rates for psychrotrophs, proteolytic psychrotrophs, lactic acid bacteria	Murray et al. 1983
	100mL/min (99,5%) for 25 min; CLOSED SYSTEM	250 mL	7°C storage for 10d	inhibitory effect of psychrotrophs(41%), Pseudomonadacese (79%), Enterobacteria (33%), protease and lipase activity (95 and 65% respectively), no effect on lactic acid bacteria	Dechemi et al.2005
	N_2 at two rates: N1, 120 mL/min;	120 mL	6,7,12°C / 6-7d	maximal effect at 6°C: significant reduction of mesophiles,	Munsch-Alatossava et al 2010a, b

Table II. Continued

N2, 40mL/min (purity of 99,9999%), continuous flushing: OPEN SYSTEM			psychrotrophs, protease and lipas producers without raise of anaerobes exclusion of PLases producers, and *B.cereus*	
N_2 flow of 4L/min (O_2 still of about 0.014%) : 6h bubbling followed by continuous flushing: OPEN SYSTEM	60-120L	5.5±0.5°C/6-7d	extension of storage life by 2.5 fold longer time compared to controls (from total counts) effect on phospholipases producers and *B. cereus*	Munsch-Alatossava et al.

Table 3. pH values of the milk from some experiments performed at laboratory and at pilot scale

Temperature °C					
	C	N1	N2	Cpilot	Npilot
6.0	6.8	7.0	6.9		
7.0	6.7	6.8	6.6		
12.0	6.4	6.6	6.2		
5.5 ±0.5				6.7 and 6.7	6,8 and 6,8

Note: The pH values of the N_2 treated milk do not disqualify the milk for further use, at both laboratory and pilot plant scales. At pilot plant scale, although no sensory analyses were performed so far, bad odours are released from the control tanks contrarily to the treated tanks (Munsch-Alatossava et al.2010b, and submitted).

Table 4. Effect of pure N_2 flushing, on bacterial groups enumerated from experiments performed at laboratory scale (a) from 3 experiments performed at 6°C and at 12°C, (b) at pilot scale from 2 experiments performed at 5.5 ± 0.5°C, determined by the differences in log values between controls (C) and treated milks ; (N1: 120 mL/min , and N2: 40 mL/min of N_2); (Munsch-Alatossava et al. 2010a,b, submitted and unpublished data)

120 mL of raw milk were flushed at laboratory scale during 6-7d and 4d for the experiments performed at 6 and 12°C, respectively

bacterial groups	6°C		12°C	
	Δ log N1-C	Δ log N2-C	Δ log N1-C	Δ log N2-C
Total aerobes	-4.5; -3.9	-3.2; -2.5	-3.4; -3.1	-2.3; -2.0
Aerobic psychrotrophs	-4.5; -4.0	-3.2; -2.4;	-3.4; -3.2	-1.5; -1.4
Aerobic protease producers	-4.8; -4.3	-3.4; -2.6	-4.1; -3.5	-2.7; -1.4
Aerobic lipase producers	-5.2; -3.1	-4.6; -2.1	-3.4; -3.3	-1.3; -0.6
Aerobic phospholipase producers	-8; -6	-3; -2	-9;-8	-9
B. cereus	-7.7;-6.9	-7.7; -3.4	ND	ND
Listeria	-4.6; -3.5	-1.8;-1.6	ND	ND
Enterobacteria	-4.2 ; -3.9	-2.4; -3.0	-3.3	-2.1
Lactobacilli	-0.7; -0.5	-0.3; +0,1	-0.7; -0.8	-0.6; -0.5
Total Anaerobes	-1.9; +0.1	-0.5 ;+ 0.4	-1.1; -0.9	-0.3; -0.1

Note: The different groups were enumerated on following media: Total and psychrotrophic aerobes/Total anerobes on PCA (Plate Count Agar); Aerobic protease, lipase and phospholipase producers on PCA+Skim milk, modified Tributyrin, PCA+Egg Yolk respectively; *B. cereus* on Mannitol Egg Yolk Polymyxin B; *Listeria* spp. on *Listeria* enrichment media; Enterobacteria on Violet Red Bile agar, Lactobacilli on MRS.

Table 4. (Continued)

b) 110L raw milk were treated at pilot scale

bacterial groups	5.5°C	
	Δ log N-C /Total[a]	Δ log N-C /Positive[b]
Total aerobes	-1.7; -1.6	
Aerobic phospholipases producers	-1.3; -0.8	-2.3; -1.8
B.cereus	-0.2; -0.2	-2.5; -2

Note: [a] corresponds to the total counts on the respective media (Plate Count Agar for Total aerobes; Plate Count Agar supplemented with Egg Yolk for the PLases producers; Mannitol Egg Yolk Polymyxin B for *Bacillus cereus*) ; [b] corresponds to the colonies that expressed the expected phenotypes (PLase positive and *B. cereus* type).

The observation that PLs producers (among them *Bacillus cereus* type) were excluded in raw milk is of major technological importance, impacting on raw milk quality as the integrity of the fat globule membrane may be preserved; but considering that different types of PLs could be pathogenic determinants (Schmiel and Miller 1999), and since the exclusion seemed not to be gram-specific this may be particularly meaningful for the raw milk´s safety. Psychrotrophic counts were kept 4-4.5 log units lower at 6°C with the high N_2 flow (N1) compared to controls (Table IV, Munsch-Alatossava et al. 2010b) contrarily to the study by Dechemi et al (2005) which indicated that among the gases and gases combinations tested, pure N_2 was the least efficient to achieve a high control of bacterial growth. More recently, we investigated the applicability of the treatment at pilot scale: N_2 gas separated from compressed air (that contained still about 0.014% O_2) was bubbled for 6h and than continuously flushed in a milk tank (where gas exchanges with the environment were still possible) ; the treatment enabled an extension of storage life of up to 110 L raw milk by 2.5 fold (Tables IV) ; PL ases producers and *B. cereus* types were not totally excluded but were still over 2-log units lower as compared to the corresponding controls (Table IV) despite the fact that the lower N_2 purity is limiting the efficiency of the flushing at pilot scale.

CONCLUSION

Additional control systems that could reinforce the cold chain, wherever it exists, or could improve the raw milk´s quality and safety wherever the cold chain fails would be indeed of value. The results obtained by flushing pure N_2 gas into raw milk containers kept as an open system (that reflect more accurately the real storage and transportation conditions of raw milk) offer an interesting perspective to target the spoilage and pathogenic potential of both psychrotrophs and mesophiles, even though many points remain unanswered. Noteworthy, the N_2 based treatments had no effect on the initial bacterial load (Munsch-Alatossava et al. 2010b). The inhibitory effects obtained for certain bacterial groups in an open system seemed to be superior to those observed for closed systems as reported by Murray et al (1983) and Dechemi et al. (2005). The results at pilot scale however further extend the potential of N_2 gas based treatments observed at laboratory scale, and constitute somehow a good starting point for practical applications at dairy farming and industrial levels (Munsch-Alatossava et al.).

The N_2 gas treatments, with a concomitant O_2 exclusion, like with CO_2 applications, also inhibit the growth of aerobes (Table IV); qualitative changes, between treated and control milks, were observed at the population level underlying N_2 treatments on Mac Conkey agar for example, where clearly lactose non-fermentors were disadvantaged by the treatments (unpublished data). Among the bacterial groups investigated so far none was really favoured by the controlled atmosphere based on 100% N_2 (Table IV, and unpublished data); the constant flushing seemed also not to favour anaerobes, or anaerobic enzyme producers (Table IV and unpublished data). From the studies performed at laboratory scale we observed that phospholipases producers, among them *Bacillus cereus* type, were "sooner or later" excluded from the raw milk, even though the N_2 purity is most probably limiting the intensity of the effect at pilot scale (Table IV).

Since phospholipids constitute the substrate of phospholipases, it is tempting to suggest that the membrane may be one primary target of N_2 action. Two hypotheses could be considered: N_2 induces direct structural changes at the membrane level, or perturbates the synthesis of essential proteins associated with biologically active membranes.

At the level of modified atmospheres, not many studies considered the biological effect of pure N_2 itself, or investigated whether N_2 amplifies an effect due to CO_2 (when both gases were simultaneously introduced); a link between the two gases has been at least established for plants *Medicago sativa* known as alfalfa, for which CO_2 fixation at the nodule level is crucial for an efficient N_2 fixation (Fischinger et al. 2010).

Data suggest that MAs in general seem safe unless storage happens at abuse temperatures where complex interactions between the natural microflora and pathogens may occur (O´Beirne 2010); but conditions, species and strain dependant effects shall be remembered: CO_2 used as a single component had little or no inhibitory effect on the growth of *E. coli* O157:H7, whereas an atmosphere composed of $O_2/CO_2/N_2$ of respectively 5/30/65% was favouring the growth of the bacterium (Abdul-Raouf et al. 1993, Diaz and Hotchkiss 1996, O´Beirne 2010). Proteolytic and non-proteolytic strains of *Clostridium botulinum* do not respond in a same way to CO_2, which had a little effect on gene expression or neurotoxin formation for one proteolytic strain (Artin et al. 2010).

More studies need to be undertaken in order to further examine the technical feasibility of the N_2 gas based treatments, to improve their efficiency, to optimise the treatments without neglecting safety aspects; thorough examination of physico-chemical, sensorial properties of treated milks needs to be undertaken, besides investigating the effects of N_2 on raw milk bacterial types, and elucidating the mechanism underlying the exclusion of some bacterial types.

ACKNOWLEDGMENTS

The authors thank Ass. Prof. O. Gursoy for his contribution to the studies performed with N_2. M. Arto Nieminen and M. Tapio Antila are gratefully acknowledged for their technical assistance in assembling the N_2 system. We thank BSc(Engin) Jyri Rekkonen for all his help to organise the raw milk delivery until the pilot plant. This review is dedicated to the memory of Me Carbiener Marthe.

REFERENCES

Abdul- Raouf UM, Beuchat LR, Ammar MS. 1993. Survival and growth of *E.coli* O157:H7 on salad vegetables. *App. Environ. Microbiol.* 59: 1999-2006.

Alvarez VB and Ji T. 2003. Emerging processing and preservation technologies for milk and dairy products. In Food Science and Food Biotechnology, Ed. Guttierrez-Lopez GF and Barbosa-Canovas GV. CRC Press: pp 313-323.

Artin I, Mason DR, Pin C, Schelin J, Peck MW, Holst E Radstrom O, Carter AT. 2010. Effects of carbon dioxide on growth of proteolytic *Clostridium botulinum*, its ability to produce neurotoxin, and its transcriptome. *App. Environ. Microbiol* 76: 1168-1172.

Ben-Yoshua S, Beaudry RM, Fishman S, Jayanty S, Mir N. 2005. Modified atmosphere packaging and controlled atmosphere storage. In Environmentally friendly technologies for agricultural produce quality. Ed Ben-Yoshua S : CRC Press: pp 61-102.

Bérard JE. 1821. Mémoire sur la maturation des fruits. *Ann. Chim. Phys.* 16 : 152-183.

Chambers JV. 2002. The microbiology of raw milk. In Dairy microbiology handbook, 3rd edition. Ed RK Robinson : pp 39-89.

Chen JH and Hotchkiss JH.1991. Effect of dissolved carbon dioxide on the growth of psychrotrophic organisms in cottage cheese. *J. Dairy Sci.* 74: 2941-2945.

Coulomb D. 2008. Refrigeration and the cold chain serving the global food industry and creating a better future: Two key IIR challenges for improving health and environment. *Trends in Food Sci. Technol.* 19: 413-417.

Cousin MA. 1982. Presence and activity of psychrotrophic microorganism in milk and dairy products: a review. *J. Food Prot.* 45: 172-207.

Daniels JA, Krishnamunthi R, Pizvi SSH.1985. A review of effects of carbon dioxide on microbial growth and food quality. *J. Food Prot.* 48: 532-537.

Dechemi S, Benjelloun H, Lebeault JM. 2005. Effect of modified atmospheres on the growth and extracellular enzyme activities of psychrotrophs in raw milk. *Engin Life Sci.* 5: 350-356.

Diaz C and Hotchkiss JH. 1996. Comparative growth of *E. coli* O157: H7, spoilage organism and shelf life of shredded iceberg lettuce stored under modified atmospheres. *J. Sci. Food Agric.* 70: 433-438.

Dixon NM and Kell DB. 1989. The inhibition by CO_2 of the growth and metabolism of micro-organisms. *J. App. Bact.* 67: 109-136.

EC Directive 92/46/EEC June 2001. Overview for microbiological criteria for food stuffs in community legislation.

FAO 2009. Milk and dairy products (updated on 23rd december 2008).

Farber JM. 1991. Microbiological aspects of modified –atmosphere packaging technology- a review. *J. Food Prot.* 54: 58-70.

Fischinger SA, Hristozkova M, Mainassara Z-A, Schulze J. 2010. Elevated CO_2 concentration around alfalfa nodules increases N_2 fixation. *J. Exp. Bot.* 61: 121-130.

Fox CW, Chrisope GL, Marshall RT. 1976. Incidence and identification of phospholipase C-producing bacteria in fresh and spoiled homogenized milk. *J. Dairy Sci.* 59:1857-1864.

Franck JF and Hassan AN. 2002. Microorganisms associated with milk. In Encyclopedia of Dairy Sciences Vol 3, Ed Roginski, Fuquay, Cox. Academic Press : pp 1786-1796.

Garcia-Gonzalez L, Geeraerd AH, Spilimbergo S, Elst K, Van Ginneken L, Debevere J, Van Impe JF, Devlieghere F. 2007. High pressure carbon dioxide inactivation of microorganism in foods: the past, the present and the future. *Int. J. Food Microbiol.* 117: 1-28.

Gill CO and Tan KH. 1979. Effect of carbon dioxide on growth of *Pseudomonas fluorescens*. *App. Environ. Microbiol* 38: 237-240.

Godfray HC, Beddington, Crute IR, Haddad L, Lawrence D, Muir JF, Pretty J, Robinson S, Thomas SM, Toulmin C. 2010. Food security: the challenge of feeding 9 billion people. *Science* 327: 812-818.

Gonzalez-Fandos E, Sanz S, Olarte C. 2000. Microbiological, physiochemical and sensory characteristics of cameros cheese packaged under modified atmosphere. *Food Microbiol* 17: 407-414.

Gorris LGM and Peppelenbos HW. 2008. Modified-atmosphere packaging of produce. In Handbook of Food Preservation 2nd Edition: Ed S Rahman, CRC Press: pp 313-333.

Gram L, Ravn L, Rasch M, Bruhn JB, Christensen AB, Givskov M. (2002). Food spoilage-interactions between food spoilage bacteria. *Int. J. Food Microbial* 78: 79-97.

Hayes PR. 1985. Food microbiology and hygiene. Elsevier, London : pp 80-139.

Hayes MC and Boor KJ. 2001. Raw milk and fluid milk products. In Applied Dairy Microbiol 2nd Ed, Ed Marth and Steel: pp 59-76.

Hong SI and Pyun YR. 2001. Membrane damage and enzyme inactivation of *Lactobacillus plantarum* by high pressure CO_2 treatment. *Int. J. Food Microbiol.* 63: 19-28.

Hotchkiss JH and Lee E. 1996. Extending shelf life of dairy products with dissolved carbon dioxide. *Eur. Dairy Mag*. 8: 1-19.

Huis int'Veld JHJ. 1996. Microbial and biochemical spoilage of foods: an overview. *Int. J. Food Microbiol.* 33: 1-18.

Hyde JA, Trzeciakowski JP, Skare JT. 2007. *Borrelia burgdorferi* alters its gene expression and antigenic profile in response to CO_2 levels. *J. Bacteriol.* 189: 437-445.

Jakobsen M. 2010. Introduction to Supplement Issue Pathogen Combat: reducing food borne disease in Europe-Control and preventing of emerging pathogens at cellular and molecular level throughout the food chain. *Int. J. Food Microbiol.* 141: S1-S3.

James SJ, James C. 2010. The food cold-chain and climate change. *Food Res. Int* 43: 1944-1956.

Jay JM, Loessner MJ, Golden DA. 2005. Modern Food Microbiology 7th Ed.Springer.

King AD and Nagel CW. 1967. Growth inhibition of a *Pseudomonas* by carbon dioxide. *J. Food Sci.* 32: 575-579.

King AD and Nagel CW. 1975. Influence of carbon dioxide upon the metabolism of *Pseudomonas aeruginosa*. *J. Food Sci.* 40: 362-366.

King JS and Mabbitt LA. 1982. Preservation of raw milk by the addition of carbon dioxide. *J. Dairy Res.* 49: 439-447.

Kisaalita WS. 2010. Milk in the cold chain. CoolChurn for smallholders. *Int Dairy Magazine* 6: 28-30.

Kuijt I andFinlayson B. 2009. Evidence for food storage and predomestication granaries 11,000 years ago in the Jordan Valley. *PNAS* 106: 10966-10970.

Leistner L and Gould GW. 2005. Update on hurdle technology approaches to food preservation. In Antimicrobials in Food, 3rd Edition. Ed Davidson, Sofos, Branen. CRC Press: pp 621-631.

Ma Y, Barbano DM, Santos M. 2003. Effect of CO_2 addition to raw milk on proteolysis and lipolysis at 4°C. *J. Dairy Sci.* 86: 1616-1631.

Mac Gowan JE. 2006. Resistance in nonfermeting gram-negative bacteria: multidrug resistance to the maximum. *Assoc for prof. in infection control and epidemiology*, 34: S29-S37.

Mac Mahon MAS. 2000. The expression of proteinases and haemolysins by *Aeromonas hydrophila* under modified atmospheres. *J. App. Microbiol.* 89: 415-422.

Marth EH. 1998. Extended shelf life refrigerated foods: Microbiological quality and safety. *Food Technol.* 52: 57-62.

Martin JD, Werner BG, Hotchkiss JH. 2003. Effects of carbon dioxide on bacterial growth parameters in milk as measured by conductivity. *J. Dairy Sci.* 86: 1932-1940.

Mazza G and Jayas DS. 2001. Controlled and modified atmosphere storage. In Food Shelf Life Stability: chemical, Biochemical and Microbiological Changes. Ed Robinson DS, Eskin NAM. CRC Press.

Mitz MA. 1979. CO_2 biodynamics: a new concept of cellular control. *J. Ther. Biol.* 80: 537-551.

Munsch-Alatossava P, Alatossava T. 2006. Phenotypic characterization of raw milk associated psychrotrophic bacteria. *Microbiol. Res.* 161: 334-346.

Munsch-Alatossava P, Alatossava T. 2007. Antibiotic resistance of raw milk-associated psychrotrophic bacteria. *Microbiol Res.* 162: 115-123.

Munsch-Alatossava P, Gursoy O, Alatossava T. 2010a. Exclusion of phospholipases (PLs)-producing bacteria in raw milk flushed with nitrogen gas (N_2). *Microbiol Res.* 165: 61-65.

Munsch-Alatossava P, Gursoy O, Alatossava T. 2010b. Potential of nitrogen gas (N_2) to control psychrotrophs and mesophiles in raw milk. *Microbiol. Res.* 165: 122-132.

Munsch-Alatossava P, Gursoy O, Alatossava T (submitted) Improved storage of cold milk by continuous flushing of N_2 gas separated from compressed air: a pilot scale study.

Murray SK, Kwan KKH, Skura BJ, Mc Kellar RC. 1983. Effect of nitrogen flushing on the production of proteinase by psychrotrophic bacteria in raw milk. *J. Food Sci.* 48: 1166-1169.

Newell DG, Koopmans M, Verhoef L, Duizer E, Aidara-Kane A, Sprong H, Opsteegh M, Langelaar M, Threfall J, Scheutz F. 2010. Food-borne diseases. The challenges of 20 years ago still persist while new ones continue to emerge. *Int. J. Food Microbiol.* 139:S3-S15.

O´Beirne D. 2010. Controlled and modified atmosphere packaging of food products. In Innovation in Food Engineering: New Techniques and Products. Ed Passos and Ribeiro, CRC Press: pp 467-479.

Pasteur L and Joubert M. 1877. Etude sur la maladie charbonneuse. *Compt Rend Hebdom des Séances de l´Acad des Sci* 84 : 900-906.

Philipps CA. 1996. Review: modified atmosphere packaging and its effects on the microbiological quality and safety of produce. *Int. J. Food Sci. Tech.* 31: 463-479.

Raghavan GSV, Vigneault C, Gariépy Y, Markarian NR, Alvo P. 2005. Refrigerated and controlled / modified atmosphere storage. In Processing Fruits Science and Technology. 2nd Ed: Ed Barrett, Somogyi, Ramaswany. CRC Press.

Rajagopal M, Werner BG, Hotchkiss JH. 2005. Low pressure CO_2 storage of raw milk: microbiological effect. J. Dairy Sci. 88: 3130-3138.Murray SK, Kwan KKH, Skura BJ, McKellar RC (1983) Effect of nitrogen flushing on the production of proteinase by psychrotrophic bacteria in raw milk. *J. Food Sci.* 48: 1166-1169.

Roberts RF and Torrey GS. 1988. Inhibition of psychrotrophic bacterial growth in refrigerated milk by addition of carbon dioxide. *J. Dairy Sci.*. 71: 52-60.

Ruas-Madiedo P, Bada-GancedoJC, Fernandez-Garcia E, Gonzalez De Llane D, De Los Reyes-Gavilan CG. 1996. Preservation of the microbiological and biochemical quality of raw milk by carbon dioxide addition: A pilot-scale study. *J. Food Prot.* 59: 502-508.

Schmiel DH, Miller VL.1999. Bacterial phospholipases and pathogenesis. *Microbes Infect.* 1: 1103-1112.

Silliker JH and Wolfe SK. 1980. Microbiological safety considerations in controlled-atmosphere storage of meats. *Food Technol.* 34: 59-63.

Stretton S and Goodman AE.1998. Carbon dioxide as a regulator of gene expression in microorganism. *Antonie Von Leeuwenhoek* 73: 79-85.

Theriault M, Uner J, Rahn A. 2004. Use of inert atmospheres in lead-free soldering. In Handbook of lead-free solder technology for microelectronic assemblies. Ed Puttlitz KJ and Stalter KA. CRC Press

Valley G and Rettger LF. 1927. The influence of carbon dioxide on bacteria. *J. Bact.* 14: 101-137.

Van Deripe. 2010. Method of use of gas nitrogen to achieve nitrogen washout from the body of mitochondria of the heart. US Patent 7661424 Issued on Feb 16/2010.

Welch RW and Mitchell PC. 2000. Food processing: a century of change. *British Med Bulletin* 56: 1-17.

Weller MA. 1959. Medical memoranda. Asphyxia with nitrogen. *British Med. J.* 28: 559.

Werner BG and Hotchkiss JH. 2002. Effect of carbon dioxide on the growth of *Bacillus cereus* spores in milk during storage. *J. Dairy Sci.* 85: 15-18.

Werner BG and Hotchkiss JH. 2006. Continuous flow nonthermal CO_2 processing: The lethal effects of subcritical and supercritical CO_2 on total microbial populations and bacterial spores in raw milk. *J. Dairy Sci.*. 89: 872-881.

Wolfe SK. 1980. Use of CO and CO_2-enriched atmospheres for meats, fish and produce. *Food Technol.* 34: 55-58.

Woolrich WR. 1944. The romance and engineering of food preservation. *Science* 11: 107-114.

INDEX

C

D

E

H

I

N

O

P

Q

R

S

T

U

V

W

Y